Stakeholders Management and Ecotourism

Stakeholders Management and Ecotourism looks at the thematic area of stakeholder management within the concept of ecotourism. It reviews the paradoxes that exist within the stakeholder relationships, ranging from building community resilience, collaboration aspects, measurement grids, product development, governance matters and managing conflict. It highlights, through its chapters, the diversity of issues as well as their possible solutions. This book will be of interest to students, practitioners as well as to faculty that do research in these areas.

The collection of chapters in this book can be used to give a theoretical underpinning to stakeholder management within ecotourism and provide a global applied perspective through the use of the case studies from an intellectual group of academics and practitioners.

The chapters in this book were originally published as a special issue of the *Journal of Ecotourism*.

Dimitrios Diamantis brings over 20 years of hospitality research and teaching experience. His areas of expertise include sustainability, ecotourism, destination management, branding and consumer behaviour in tourism and hospitality. He is currently an Executive Academic Dean at Les Roches, Global Hospitality Management, Switzerland.

Stakeholders Management and Ecotourism

Edited by
Dimitrios Diamantis

Routledge
Taylor & Francis Group
LONDON AND NEW YORK

First published 2022
by Routledge
2 Park Square, Milton Park, Abingdon, Oxon, OX14 4RN

and by Routledge
605 Third Avenue, New York, NY 10158

Routledge is an imprint of the Taylor & Francis Group, an informa business

British Library Cataloguing-in-Publication Data
A catalogue record for this book is available from the British Library

ISBN13: 978-0-367-75635-2 (hbk)
ISBN13: 978-0-367-75637-6 (pbk)
ISBN13: 978-1-003-16330-5 (ebk)

Typeset in Minion Pro
by codeMantra

Publisher's Note
The publisher accepts responsibility for any inconsistencies that may have arisen during the conversion of this book from journal articles to book chapters, namely the inclusion of journal terminology.

Disclaimer
Every effort has been made to contact copyright holders for their permission to reprint material in this book. The publishers would be grateful to hear from any copyright holder who is not here acknowledged and will undertake to rectify any errors or omissions in future editions of this book.

Contents

Citation Information

The chapters in this book were originally published in the *Journal of Ecotourism*, volume 17, issue 3 (2018). When citing this material, please use the original page numbering for each article, as follows:

Introduction
Stakeholder ecotourism management: exchanges, coordination's and adaptations
Dimitrios Diamantis
Journal of Ecotourism, volume 17, issue 3 (2018) pp. 203–205

Chapter 1
Learning from the locals: the role of stakeholder engagement in building tourism and community resilience
Joanne Pyke, Alexandra Law, Min Jiang and Terry de Lacy
Journal of Ecotourism, volume 17, issue 3 (2018) pp. 206–219

Chapter 2
World heritage listing as a catalyst for collaboration: can Mount Fuji's trail signs point the way for Japan's multi-purpose national parks?
Thomas Jones, Sue Beeton and Malcolm Cooper
Journal of Ecotourism, volume 17, issue 3 (2018) pp. 220–238

Chapter 3
Repertory grids and the measurement of levels of community support for rural ecotourism development
Stephen Schweinsberg, Simon Darcy and Stephen L. Wearing
Journal of Ecotourism, volume 17, issue 3 (2018) pp. 239–251

Chapter 4
Community-based ecotourism: beyond authenticity and the commodification of local people
Nantira Pookhao Sonjai, Robyn Bushell, Mary Hawkins and Russell Staiff
Journal of Ecotourism, volume 17, issue 3 (2018) pp. 252–267

Chapter 5
Surfing tourism and local stakeholder collaboration
Nick Towner
Journal of Ecotourism, volume 17, issue 3 (2018) pp. 268–286

Chapter 6

Envisioning Eden: the manufactured ecotourism environment of Singapore
Aaron Tham
Journal of Ecotourism, volume 17, issue 3 (2018) pp. 287–305

Chapter 7

Authenticating eco-cultural tourism in Kazakhstan: a supply side perspective
Guillaume Tiberghien, Hamish Bremner and Simon Milne
Journal of Ecotourism, volume 17, issue 3 (2018) pp. 306–319

Chapter 8

Governance and local participation in ecotourism: community-level ecotourism stakeholders in Chiang Rai province, Thailand
Nicola J. Palmer and Nipon Chuamuangphan
Journal of Ecotourism, volume 17, issue 3 (2018) pp. 320–337

Chapter 9

Sustainable ecotourism management and visitor experiences: managing conflicting perspectives in Rocky Mountain National Park, USA
Tek B. Dangi and William J. Gribb
Journal of Ecotourism, volume 17, issue 3 (2018) pp. 338–358

Contributors

Sue Beeton William Angliss Institute, Melbourne, Australia.

Hamish Bremner Tourism and Hospitality Department, Auckland University of Technology, New Zealand.

Robyn Bushell School of Social Sciences and Psychology, Western Sydney University, Australia.

Nipon Chuamuangphan Nakhon Pathom Rajabhat University, Nakhon Pathom, Thailand.

Malcolm Cooper College of Asia Pacific Studies, Ritsumeikan Asia Pacific University, Beppu, Japan.

Tek B. Dangi Department of Recreation, Park and Tourism Sciences, Texas A&M University, College Station, USA.

Simon Darcy UTS Business School, University of Technology Sydney, Australia.

Terry de Lacy Institute for Sustainable Industries and Liveable Cities (ISILC), Victoria University, Melbourne, Australia.

Dimitrios Diamantis Executive Academic Dean, Les Roches, Global Hospitality Education, Bluche/Crans Montana, Switzerland.

William J. Gribb Department of Geography, University of Wyoming, Laramie, USA.

Mary Hawkins School of Social Sciences and Psychology, Western Sydney University, Australia.

Min Jiang Institute for Sustainable Industries and Liveable Cities (ISILC), Victoria University, Melbourne, Australia.

Thomas Jones College of Asia Pacific Studies, Ritsumeikan Asia Pacific University, Beppu, Japan.

Alexandra Law Institute for Sustainable Industries and Liveable Cities (ISILC), Victoria University, Melbourne, Australia.

Simon Milne Tourism and Hospitality Department, Auckland University of Technology, New Zealand.

Nicola J. Palmer Sheffield Hallam University, UK.

Joanne Pyke Institute for Sustainable Industries and Liveable Cities (ISILC), Victoria University, Melbourne, Australia.

Stephen Schweinsberg UTS Business School, University of Technology Sydney, Australia.

Nantira Pookhao Sonjai Department of Hotel and Tourism Studies, Siam University, Bangkok, Thailand.

Russell Staiff School of Social Sciences and Psychology, Western Sydney University, Australia.

Aaron Tham School of Business, University of the Sunshine Coast, Sippy Downs, QLD, Australia.

Guillaume Tiberghien School of Interdisciplinary Studies, University of Glasgow, Dumfries, Scotland.

Nick Towner Tourism Hospitality, Auckland Institute of Studies, New Zealand.

Stephen L. Wearing UTS Business School, University of Technology Sydney, Australia.

INTRODUCTION

Stakeholder ecotourism management: exchanges, coordination's and adaptations

Dimitrios Diamantis

Ecotourism presents itself with various opportunities, yet the management of tourism in protected and/or unprotected areas represents great challenges (Ullenberg, 2015, p. 34) one of which is the concept of stakeholder management.

According to Backman and Munanura (2015), stakeholders are key actors to achieve sustainable development through ecotourism (p. 98). Marulo (2012) agrees by demonstrating that its management affects the growth of ecotourism, and all stakeholders form the basis on which ecotourism stands and its success is dependent on the cooperation, communication and involvement of different stakeholders (p. 50). Byrd (2007) mirrors this idea by emphasizing the need for ecotourism to be developed, planned, and managed in a sustainable way whilst integrating all stakeholders throughout the entire process (p. 8).

Since Edward Freeman, in 1984, defined the term stakeholders as key groups or individuals that can affect or be affected by a company's activities, there have been ongoing discussions and modifications in relation to who stakeholders are (Agüera, 2013).

Holistically speaking, the term stakeholder was redefined to a – 'term as a series of voluntary agreements between adults who voluntarily affirm their commitment to the objectives and purposes of the corporation, as well as entering cooperative agreements with other individuals or groups' (Byrd, 2007, p. 10). Examples of major ecotourism stakeholders that should be involved are as follows: tourists, suppliers, local governments, accommodation, tourism agencies, local communities, NGOs, ecological groups, to name just a few … (Dologlou, 2016, p. 20).

Besides stakeholders' multifaceted definition and structures, Garrod (2003) states that the stakeholder theory has proved to be an efficient ecotourism management tool as it allows for the identification of key players and emphasizes the need of involvement (p. 15). Yet, the issue presented is that while stakeholder involvement is vital for ecotourism development, decisions are made "top down" by superiors/experts and local's see this as not reflecting the opinions, and interest of their community (Hardy & Beeton, 2001, p. 168).

Kiper (2013) adds to this and explains that this occurs because there is an issue of competing interests between different stakeholders during the decision making process (p. 862). This is due to the complexity of ecotourism and the multidimensional aspect of stakeholders: as they are diverse, with different perspectives and varying interest (Darley & Moswete, 2012).

To shed a light into the above mentioned debates, this special issue brings out the following papers:

(1) The article by Pyke, Law, Jiang and de Lacy examines the stakeholder engagement in constructive resilience to the impacts of bushfire in Austrailia;
(2) The article by Jones, Beeton and Cooper brings up the complexities in stakeholder management at Mount Fuji in Japan;
(3) The article by Schweinsberg, Darcy and Wearing explores the merits of employing Personal Construct Theory based repertory grids as a methodological tool to understand the subjective realities of local people within the ecotourism spectrum;
(4) The article by Pookhao discusses articulates the connection between policy, tourism stakeholders and practices in relation to the commodification and authenticity of community based ecourism in Thailand;
(5) The article by Towner points to the current stakeholder collaboration in the Mentawai Islands surfing tourism industry by exploring multiple stakeholder perspectives;
(6) The article by Tham points to the directions of planning and stakeholder partnership as a case in Singapore for new destinations embarking on their ecotourism journey to 'Envision Eden';
(7) The article by Tiberghien, Bremner and Milne illustrates the various aspects of the commodification of nomadic culture in a post-Soviet heritage and details the role authenticity plays in the planning and development of Kazakhstani tourism and local community participation and empowerment through stakeholder management;
(8) The article by Palmer and Chuamuangphan points to *the actors' interactions as* they relate to the use of resources and environment in the Chiang Rai province in Thailand. The factor of Involvement, among other issues, in community based ecotourism appears to be linked to an individual's social standing or general position in society where is illustrated through a range of qualitative type of explorations; and
(9) The article by Dangi and Gribb emphasizes partnership and collaboration among various stakeholders for resource sustainability and visitor satisfaction in Rocky Mountain National Park, USA.

With the above collections, the special issue aims to bring into the literature of stakeholder management a fresh look into its complexities, and its paradoxes. Consequently, it aims to highlight that ecotourism is complex due to the large range of stakeholders with different interests. Perhaps, it is instructive to look at such complexities as a way to manage ecotourism more effectively through an exchange between the different stakeholders, a coordination of their action and adaptation processes. The latter mechanisms can be clearly illustrated through the special issue which in effect aims to broaden the debate and to instill the importance of stakeholder management as a way of dealing with ecotourism challenges.

Disclosure statement

No potential conflict of interest was reported by the authors.

References

Agüera, F. O. (2013). *STAKEHOLDER THEORY AS A MODEL FOR SUSTAINABLE DEVELOPMENT IN ECOTOURISM.* University of UTESA, Master in Business Administration. Seville: University of Seville.

Backman, K. F., & Munanura, I. (2015). Introduction to the special issues on ecotourism in Africa over the past 30 years. *Journal of Ecotourism*, *14*(2–3), 95–98.

Byrd, E. (2007). Stakeholders in sustainable tourism development and their roles: Applying stakeholder theory to sustainable tourism development. *Tourism Review*, *62*(2), 6–13.

Darley, W. K., & Moswete, N. N. (2012). Current issues in method and practice. Tourism survey research in sub-Saharan Africa: Problems and challenges. *Current Issues in Tourism*, *15*(4), 369–383.

Dologlou, N. (2016). *Ecotourism in Protected Areas, A Literature Review*, 1–23. National Technical University, Metsovion Interdisciplinary Research Centre. Athens: EcoClub.

Garrod, B. (2003). Local participation in the planning and management of ecotourism: A revised model approach. *Journal of Ecotourism*, *2*(1), 33–53.

Hardy, A. L., & Beeton, R. (2001). Sustainable tourism or maintainable tourism: Managing resources for more than average outcomes. *Journal of Sustainable Tourism*, *9*(3), 168–192.

Kiper, T. (2013). Role of ecotourism in sustainable development. In M. Ozyavuz (Ed.), *Advances in landscape architecture* (pp. 773–870). Istanbul: InTech.

Marulo, A. M. (2012). *Turismo e o Meio Ambiente: Uma analise do Ecoturismo e a sua Contribuicao Socio-Ambiental no Distrito de Maturtuine: Caso de Reserva Especial de Maputo-Mozambique*, 13–55 (*Masters Dissertation*). Universidade do Rio, Brazil.

Ullenberg, A. (2015). Community-Based Tourism Development and Fire Management in Conservation Areas of the SADC Region. *Evaluating Cross-Border Natural Resource Management Projects, 31–39.* Germany: SLE PUBLICATION.

Learning from the locals: the role of stakeholder engagement in building tourism and community resilience

Joanne Pyke, Alexandra Law, Min Jiang and Terry de Lacy

ABSTRACT

This article explores the role of stakeholder engagement in building resilience to the impacts of bushfire. The discussion is informed by a study of Harrietville – a nature-based tourism destination in NE Victoria, Australia, that was affected by a major bushfire in 2013. Harrietville is a small town that acts as the gateway to premier ski fields and provides access to abundant nature-based attractions including iconic bushwalks, touring routes and rivers. The research aim was to investigate the impact of bushfire on the tourism economy and was based on a visitor survey and in-depth interviews with representatives from across stakeholder groups. A Destination Sustainability Framework (DSF) was applied to identify key vulnerability and resilience characteristics of the town as well as to identify opportunities for adaptation to future fires. Fragmented stakeholder communications and a lack of community engagement in planning for bushfire prevention, emergency management and recovery are identified as a key vulnerability. In response, government has successfully implemented initiatives to strengthen stakeholder relationships and engage the community in emergency planning processes. The article provides an important case study of resilience building by small tourism destinations as well as highlighting the importance of connecting with local knowledge in bushfire management.

Introduction

Throughout Australia, many small regional economies are dependent on nature based tourism. As a result of global warming, however, the increasing frequency and severity of bushfire represents a major threat to communities reliant on incomes and employment generated by visitors drawn by nature-based attractions. The resilience of such destinations, and the communities that tourism spending supports, is therefore deeply reliant on sympathetic planning and actions of multiple land managers and emergency services responsible for fire prevention, emergency response and recovery.

This paper draws on the findings of a study conducted with the aim to assess the economic value and vulnerability of nature based tourism of a destination affected by four bushfire events since 2003. The case-study destination is Harrietville – a small town based in the Australian Alpine Region located on the Great Alpine Road between the regional centre of Bright and Mt Hotham, one of Victoria's premier ski resorts. Despite its small population of around 400 residents, the town provides immediate access to abundant natural attractions including a national park and state forest, rivers, ski-fields and iconic touring routes for walkers, cars, motor-bikes and cyclists. One implication of this diversity of resources is that the tourism system and governance processes are highly complex despite the smallness of the destination. Key stakeholders include tourism operators, multiple land management authorities, emergency services and social and community services. The local residential community is also enmeshed with both the tourism system and governance bodies both directly through property, employment and/or business interests, and indirectly through community networks and an investment in the character of the locality.

The impetus for this study emerged in the context of the most recent bushfire, which started on 21 January 2013 and was ignited through a lightning strike at Smoko, a locality around four kilometres north-east of Harrietville. While the fire was named 'the Harrietville Fire', the fire actually travelled around Harrietville to within 100 metres of Mt Hotham and across the Great Alpine Road. The result was that 37,000 hectares of Alpine National Park were burnt over a period of 55 days leading to the tragic death of two firefighters. The fire was also followed by heavy rains, which coupled with the after-effects of the fires, caused considerable flooding and water contamination in the township. The combined effects of fire and flood also caused mud slides forcing road closures between Harrietville and Mt Hotham. These events resulted in at least partial road closures to Harrietville for twelve weeks during 2013. As a result, the town suffered a loss of income from visitors or transit tourists for effectively three months of 2013. Further, damage was caused to parkland and many walking and four-wheel drive trails through parkland were either damaged, closed for safety reasons or difficult to access. A compounding issue was the reputational damage caused by misleading media coverage of the fire events.

The impact of the fire prompted a multi-agency government response primarily led by the state Emergency Management Authority. A key strategy was the support and facilitation of a community forum comprised of representatives of tourism operators, general community members and community groups and services such as the Country Fire Authority (CFA). The forum provided the communication mechanism for a number of projects designed to support Harrietville's planning, emergency response and recovery for future fire events. The study of tourism values in the area was part of this effort and one of the key resilience factors identified was the success of stakeholder engagement efforts, and the positive impacts this has made to community preparedness and response to fire events. At the same time, the process had been undertaken from an emergency management perspective and tourism values of the area have been a secondary, rather than a central focus of stakeholder engagement efforts. By taking a perspective focussed on the tourism economy, the study revealed important gaps in stakeholder recognition and engagement that have negatively impacted on both the integrity of the nature-based resources on which tourism relies and consequently, the resilience of the area as a tourism destination. Visitors, future visitors and to a lesser extent, tourism operators, have been particularly overlooked as a stakeholder in the process of fire prevention and recovery.

Informed by a Destination Sustainability Framework (DSF) (Calgaro, Lloyd, & Dominey-Howes, 2014), this paper presents the key vulnerabilities and resilience characteristics identified through the tourism values study. Drawing on stakeholder theory (Bourne, 2008; Byrd, 2007; Sautter & Leisen, 1999), the role of stakeholder engagement and communications in bushfire prevention, management and recovery is discussed. In particular, the intersections between emergency management authorities and the local tourism economy are highlighted.

The tourism resilience and stakeholder nexus

Harrietville is surrounded by expanses of bushland and bushfire risk is accepted as a part of life. However, the frequency and severity of bushfires in Australia are expected to increase (Hughes & Steffen, 2013) and in recent years, several major bushfires events have occurred in the Harrietville region in close proximity. These included the bushfire events of 2003, 2006/07, 2009 and 2013.

Traditionally risk analysis (see for example, ISO Guide 73, 2009) has been a top down approach identifying individual risks and estimating the likelihood of the risk occurring and the severity of its impact should it occur, to gain a quantitative, if relative, estimate of risk. More recently a more bottom up approach has come to the fore based around estimating a communities' vulnerability to a broad range of risks. To account for this change, this study approached the vulnerability assessment from a climate adaptation perspective where 'adaptation' refers to adaptive actions to reduce vulnerability and increase resilience to the potential adverse impacts of climate extremes (IPCC, 2012).

Tourism has unique challenges and opportunities in regards to climate adaptation. Not only is tourism demand sensitive to changes in climate (for example skiing destinations that rely heavily on a conducive climate), climatic changes also pose risks to the appeal of a destination, transport infrastructure, the resource base (natural and human), tourist satisfaction, safety and health, as well as the sustainability of tourist facilities (Jopp, De Lacy, & Mair, 2010). The adaptive capacity also varies between tourism stakeholders (Scott et al., 2008). For example, accommodation providers, attraction operators and communities generally have a low adaptive capacity as they are bound to their geographic area, whilst their customers (tourists) are mobile and can simply choose a different destination to suit their individual needs.

To address individual climate change risks, a wide range of adaptations are used by tourism stakeholders around the world, including technological, managerial, policy and behavioural adaptation (Scott et al., 2008) and a number of adaptation frameworks have been proposed in the literature. These include the Regional Adaptation Framework (Jopp et al., 2010; Jopp, De Lacy, Mair, & Fluker, 2013), the vulnerability assessment methodology of Moreno and Becken (2009), the vulnerability scoping diagrams (VSD) of Polsky, Neff, and Yarnal (2007) and the Climate Change Vulnerability/Resilience Framework of Klint (2012). For this study, the Destination Sustainability Framework (Calgaro, Dominey-Howes, & Lloyd, 2014; Calgaro, Lloyd et al., 2014) was selected as the most appropriate methodology as it was developed in the context of a natural disaster (the Tsunami of 2006) and has since been applied in adapted forms in a number of destinations in the South Pacific including Samoa and Vanuatu (Jiang & DeLacy, 2014; Klint et al., 2012). The framework focuses on exposure, sensitivity and adaptive capacity as the key elements of vulnerability and incorporates feedback loops that highlight the inter-dependencies of the drivers within the system.

Whilst not an explicit part of the model, the DSF illuminates the role that stakeholder engagement plays in the vulnerability and resilience of a destination (Calgaro, Dominey-Howes et al., 2014; Calgaro, Lloyd et al.,Lloyd, 2014). As such, the approach has commonalities with the literature highlighting the importance of stakeholders in achieving sustainability (e.g. Waligo, Clarke, & Hawkins, 2013). Identifying who the stakeholders are, and how to engage with them, plays a critical role in resilience building of a destination.

Stakeholder theory has its roots in the works of Freeman (1984, p. 4), who identified stakeholders as 'any group or individual who can affect or is affected by the achievement of the organisation's objectives.' Since then, the literature has evolved and justified the concept in a variety of contexts, which has included sustainable tourism (e.g. Bramwell & Sharman, 1999; Byrd, 2007; Fletcher, 2009; Sautter & Leisen, 1999; Waligo et al., 2013) and disaster risk reduction (e.g. Gaillard & Mercer, 2013; Mercer, Gaillard et al., 2012; Mercer, Kelman, Taranis, & Suchet-Pearson, 2010; Palttala, Boano, Lund, & Vos, 2012). Tourism, in its myriad of forms, provides a unique stakeholder context as it involves people travelling to a destination outside their usual environment. This gives tourism destinations a complex multi-stakeholder nature and the literature often identifies six tourism stakeholder types: tourists, industry, local community, government, special interest groups and educational institutions (e.g. Waligo et al., 2013). However, in the context of vulnerability and resilience of a destination, stakeholders also include the emergency and recovery services and others depending on the context. Culture and traditional knowledge, for instance, have been shown to play a critical role in disaster risk reduction and the vulnerability of a destination (Bird, Gísladóttir, & Dominey-Howes, 2011; Gaillard & Mercer, 2013; Mercer et al., 2012; Mercer, Kelman et al., 2010). However, effective stakeholder collaboration between each of these groups can be complex and very difficult to achieve. Different stakeholders often have disparate interests and/or perspectives, while the engagement process is influenced by the quality of leadership, quality of information and accessibility, stakeholder involvement capacity, stakeholder relationships and implementation priorities (Waligo et al., 2013). This is particularly the case in the tourism context where the mix of stakeholder groups is highly heterogeneous with diverse interests.

Method

Informed by the DSF (Calgaro, Lloyd et al., 2014), a vulnerability/resilience (V/R) assessment of tourism in Harrietville was conducted to reveal the risks (shocks and stressors) for Harrietville's tourism system in the context of bushfire, the constraining factors increasing its vulnerability, and the enabling factors increasing its resilience. A mixed method approach was adopted in order to gather relevant data, both primary and secondary, to undertake the V/R analysis. Specific data collection methods included visitor survey and in-depth interviews with relevant stakeholders.

A survey of visitors to Harrietville was conducted over the 2014 Easter period. With reference to relevant literature, (Sanders, Laing, & Houghton, 2008; Tourism Research Australia, 2014), the survey included 22 questions across three sections: purpose and motivations for visiting Harrietville, understanding of bushfire risks and impacts, and demographic details. A total of 285 usable surveys were completed and analysed using SPSS. The survey findings were used to inform the V/R analysis which considers visitors as an essential component of the tourism system and a key tourism stakeholder. Visitors' are potentially exposed to bushfire risks, and their understanding of and preparedness for bushfire risks and impacts have significant implications for determining a tourism destination's level of vulnerability and resilience.

A total of 25 semi-structured face-to-face in-depth interviews were also conducted with a range of tourism stakeholders, who were identified to ensure representation by a cross-section of people connected with the Harrietville tourism system. Major stakeholder groups interviewed included: private tourism businesses, community members, relevant government agencies at state, regional, and local levels, and other peak tourism and event management organisations. The interview data was analysed to identify categorical themes (Patton, 2002) reflecting the detailed components featured in the (Calgaro, Lloyd et al., 2014).

The DSF provides a conceptual framework to understand *who* is vulnerable or resilient, to *what* and *why* (Calgaro, 2010). The process of analysis revealed how different stakeholders in the tourism system collectively contribute to economic, social, environmental, legal and political conditions that determines vulnerability and resilience. In the context of bushfire risk, it is crucial to understand how different stakeholders interact with each other in the complex governance processes related to tourism development, natural resources management, disaster preparedness and reduction, and emergency management. To this end, the V/R assessment highlighted the role of stakeholder engagement and communications in building resilient tourism destinations. With a particular focus on stakeholder communications, the following section describes key themes to emerge from the interviews and survey findings in relation to the V/R categories of analysis.

Key findings

Shocks and stressors

As indicated, the first stage of a V/R analysis is the identification of shocks (immediate impacts) and stressors (slow onset impacts). In this case, the key shock was a severe fire that caused extensive damage to 37,000 of national parkland over a period of 55 days as well as leading to the tragic death of two firefighters (Hallowes, 2013). Immediately following the fire was a period of flooding rain causing further shocks to the system. With little vegetation left, burnt mountain areas are particularly susceptible to erosion, when the 'first lot of rain sends all the silt, all the ash, all the soot, everything else down and kills fish, kills whatever wildlife is left' as one of the respondents pointed out. The heavy rains consequently resulted in landslides, one of which severed the only road connecting Harrietville with Mt Hotham. The Ovens River, already clogged with post-fire debris became completely blocked, eventually flooding parts of the town. Some properties and tourism businesses were badly affected, including washed-away bridges and footpaths, as well as flooded rooms and destroyed gardens – requiring what are effectively major clean-up operations for small businesses with only a few employees and affecting their ability to trade as normal. The ash and debris run-off caused considerable water pollution with impacts on the visual amenity (e.g. dead fish and platypus caught in sticks) and death to fish stocks.

A number of additional stressors, however, were identified as exacerbating the impacts of the fires. A key issue was sensationalised and misleading news coverage which continued throughout the course of the fire, and then immediately stopped when the fire had stopped with little public communications to explain that Harrietville was not damaged and that it was safe to return. A particular source of community frustration was that the fire had been named the 'Harrietville Fire', thereby impacting tourism visitation even when it was safe to return and that the fire had not actually affected the township.

Further stress was placed on the town due to fire-fighting operations based in Harrietville as well as road and track closures that blocked access to the town and surrounding bushland. While such actions were understood as unavoidable and necessary to both contain the fire and to ensure public safety, they were nonetheless part of the wider context that impacted on the local community and tourism system. This was particularly the case given that fire operations continued for over six weeks, and local road closures continued for ten. Given that access to the town was so limited, so too was any tourism spending. As one respondent said, '... so really, from the 24th January to Easter, the town basically had no income'. Road works also limited access for the myriad of other services a tourism business relies on, such as the delivery of goods, as well as repair and maintenance services, often conducted at that time of year. A further effect was a loss of work for those in the community that work at Mt Hotham. The stress of road closure was compounded by the use of road signage that was misleading, limiting the existing access even further as it did not give exact details of closure points, opening times or reasons how and why it was closed. Signage was easily interpreted as if it was not possible to reach Harrietville when in fact it was – thereby confusing potential visitors even further. Finally, repeated road closures over the long term may become an image problem for the destination, as '... there is a fear that visitors will start to see Harrietville as unreliable and travel to Hotham via Omeo'.

Other key related stressors include those related to climate change. The increasing incidences of summer heat waves act as a deterrent to visitors in itself as well as create the conditions for fires. A related stressor is unreliable snow seasons where poor snowfalls in the ski fields further discourage winter visitation. However, climate change is an underlying factor in each of the identified, bushfire-related shocks and stressors. Fire seasons are expected to lengthen into the future, further reducing the opportunities for safe hazard reduction burning (Hughes & Steffen, 2013).

Vulnerability

Vulnerability can be defined as 'the degree to which an exposure unit [households, human groups, ecosystems, and communities] is susceptible to harm due to exposure to a stress, and the ability (or lack thereof) of the exposure unit to cope, recover, or fundamentally adapt' (Kasperson, Turner, Schiller, & Hsieh, 2001, p. 7). It is a property of the coupled human-environment system that is determined by three interconnected dimensions: exposure, sensitivity and system adaptiveness (Calgaro, Lloyd et al., 2014).

On each of these dimensions, Harrietville can be identified as being vulnerable to the incidence of shocks such as a fire. First, the town is highly exposed with bushfire risk being generated by steep terrain and heavy vegetation on all sides (CFA, 2014, p. 2). Heavy vegetation and topography create the potential for severe ember attack and direct fire into the town environs with properties located higher in the valley being particularly vulnerable due to the tendency of fire to travel rapidly up-hill.

While tourism offerings have become increasingly diversified in recent years, visitation is primarily driven by the nature-based assets in the area. As identified by the visitor survey, bushwalking, picnicking, driving tours, cycling and fishing are the main activities undertaken by visitors. In the absence of other industry besides forestry, Harrietville's economy is economically dependent on nature-based tourism as well as tourism transit (primarily to Mt Hotham but also as part of a number of car, bike, walking and four wheel drive routes). All of Harrietville's tourism activity can therefore be affected by bushfire for extended periods of time – and long past the time when emergency response operations have been completed. After the 2013 bushfire, tourism revenue was significantly affected for several months (Pyke, Jiang, de Lacy, Whitelaw, & Jones, 2015).

The environment is also highly sensitive. The succession of intense fires approximately every three years since 2003 has, in particular, had severe consequences for a number of plant species. The most substantial impact has been on the regeneration of the tall Alpine Ash (Eucalyptus delegatensis), a species that is an 'obligate seeder' and needs to mature before it can regenerate. Large areas of Alpine Ash were lost in 2013 and attempts to re-seed have been only partially successful. This is a particular concern from a tourism perspective as it is the bushland that is the major asset for Harrietville's tourism system.

As a small town with few services and overshadowed by the 'hero destinations' (NE Victoria Tourism Board, 2012) of Mt Hotham to the south and Bright to the north, Harrietville has a particularly sensitive image. This sensitivity was particularly highlighted by the effects of media coverage surrounding the fire. According to the visitor survey, more than 60 per cent of tourists learnt about fires thought general media, which tended to sensationalise the fire events, and may have altered tourist's choices and preferences of destinations. One interviewee expressed the concern about the effect of this on future visitation by saying,

> I guess the other area we don't know is, is it changing people's perceptions on visiting not only Harrietville but anywhere in the bush in the summer time? Is that perception changing? I don't know but we do know that we were having cancellations right through the year based on the fact that they thought that things would be different [after the fire].

Tourism development and services in the town is also limited and vulnerable due to gaps in internet coverage and the lack of a town sewerage system which limits building development due to planning restrictions. The greatest vulnerability, however, is that there is only one road through the town. If this is closed, as it was in 2013, there is no visitor access. Overall, while the smallness of the town, and its low-key image contribute to the appeal of the area to visitors, limited services, communications infrastructure and gaps in services, means that the town is highly sensitive to negative visitor perceptions which may or may not be informed by accurate information.

One of the primary vulnerabilities that were exposed by the 2013 fire emerged due to poor communications between governance agencies and the community. As discussed, responsibility for land management for the parks, emergency management services and the town lies with multiple organisations and agencies. Interview results revealed some on-going concerns by the community and private sector in relation to a perceived lack of co-ordination, communication and community engagement by government agencies concerning land and fire risk management. For example, interviewees sited instances where four wheel drive tracks have been closed without community consultation and such measures have impacts on visitors and business operators. At the same time, planning, governance and land management officers highlight the challenges of implementing national or state policy within the area, with limited human and financial resources, and in a large and geographically diverse region. Concerns were also expressed by government officers about the difficulties in managing some inherent conflicts between the necessary implementation of safety or conservation measures and the interests of varied tourism business operations that rely on nature-based tourism assets. The effects of these communication gaps were also exacerbated by the social and emotional strain of living in a bushfire-risk area where tensions run high over extended periods and variously impact on different sectors within the community. As one community member commented,

> I think that is something that is undermined, that the communities that live up here are often so heavily involved [ ] in the community support networks .... You can literally go weeks and weeks and weeks and weeks in these communities up here and it's just women in town or its single mums all struggling because husbands are away fighting fires. [ ] That has a massive impact on people's ability to be resilient and to think clearly. They're exhausted ...

A further communication gap was revealed in relation to fire awareness and preparedness. A concern of local government officers and land management agencies was community complacency in relation to fire planning. While awareness was heightened immediately following a fire, over the longer term, few tourism businesses are actively planning for the expected increases in risk and continue as 'business-as-usual'. This has implications for visitors whose primary source of information about local conditions is through tourism operators. Without a well thought out visitor communications strategy, many visitors are vulnerable. This is particularly the case for first-time visitors and those who stay in holiday homes making them harder to reach. Visitor communications is particularly important given visitor survey results which showed that more than half of the visitors surveyed (51.8%), were not aware of any information about what to do in case of a bushfire in the area.

Combined, the research findings showed that the tourism system in Harrietville is exposed to bushfire risk, is sensitive to a number of effects from fire, and governance, structural and communications issues work to prevent comprehensive adaptation to the increasing frequency and intensity of fires. At the same time, the events of 2013 were not overwhelming due to some key strengths or resilience factors that support the recovery of tourism following the shock of bushfire and its after-effects.

Resilience

Through the lens of the DSF, resilience is understood as 'the ability of a social or ecological system to absorb disturbances while retaining the same basic structure and ways of functioning, the capacity for self-organisation, and the capacity to adapt to stress and change' (IPCC, 2007, p. 880) and is a direct expression of the strength of a coupled human-environment system (Carpenter, Walker, Anderies, & Abel, 2001). Harrietville exhibits a number of characteristics that enable the town to absorb and adapt to shocks. Much of this resilience can be attributed to strengths in the tourism system, community resources and governance arrangements that have adapted by making efforts to strengthen stakeholder relationships and communications systems.

A key resilience factor is that there are multiple nature-based attractions and activities accessible from Harrietville, low seasonality, an increasingly diversified visitor base and high rates of repeat visitation. As revealed by the visitor survey, visitors engage in a wide range of activities, appreciate the area for its beauty and relaxing atmosphere, and a high proportion (74.6%) had visited Harrietville at least once before (Pyke et al., 2015). While traditionally, the main reasons for visitation was for skiing at Mt Hotham in winter, or bushwalking in the warmer months, the motivations for visiting have expanded with the popularity for the area for food and wine experiences, cycling and four-wheel driving (Alpine Shire Council, 2013). This growth has been promoted through strong tourism planning at a regional and local government level.

One of Harrietville's key strengths is its community that was widely described as politically savvy, organised, tenacious and well-resourced. As identified by a recent regional business survey (Alpine Shire Council, 2011), the key motivation for operating a business in the area is the lifestyle that it offers and there is a strong community pride in being self-sufficient and community spirit is deeply valued. Given the exposure to bushfire risk, residents are also accepting of the need to adapt to the conditions. As one interviewee commented, '… you can't just ring the CFA, they might be busy and the ambulance is half an hour away.' There is a clear sense within the community of needing to accept the realities of fire, and to '… just pull your socks up, get stuck in and just keep going.' The community is concerned but not afraid of future fires and is strongly immersed with community organisations and initiatives.

A third key resilience factor is trust in, and engagement with emergency response processes and agencies. For instance, there is great community trust in the Harrietville CFA. There is also considerable appreciation of the role of local, state and regional agencies. For example, the risk of water contamination during the 2013 floods was averted through the provision of portable water and chlorine filter treatment. The perceived strength of Victoria's overall emergency response system provides a platform on which resilience for tourism can be built. Until recently, tourism was not a priority in immediate emergency response but awareness is growing and new systems and processes are being put in place with the establishment of a community forum that provides a mechanism for improved communication between the community and emergency management agencies. As one community member commented,

> … in terms of the emergency response side of things for bushfire suppression, I'd say the planning has been quite adequate but as we talked about before, it's about bringing the community on with them.

While this is an ongoing process, immediate emergency responses are widely recognised as highly successful in the protection of life and property. Both the system in itself, as well as community trust in this system, is an important dimension of the capacity of the system to adapt to future fire risks.

Further to the emergency response planning, there have also been adaptations with the aim to improve relationships, communications and engagement for all aspects of emergency planning, management and recovery. A Community Emergency Management Plan (CEMP) for Harrietville, for instance, has been developed by Victoria's fire agencies, emergency services partners and the community. It aims to support people in Harrietville to be better prepared and manage the risks from bushfire, flood and landslide events before, during and after emergencies. This project was a key strategy to build government and community relationships through collaborative emergency management planning.

Overall, Harrietville can be identified as having strong capacity for adaptation yet the underlying stressor of climate change will mean that resilience capacity will be increasingly tested with the predicted increase in fire intensity and frequency. As a consequence, and in the interests of long term sustainability, the vulnerabilities revealed and lessons learned through the fires need to be addressed.

Discussion

Through undertaking a V/R analysis, a number of key factors that limit the adaptive capacity of the tourism economy were revealed. Fragmented stakeholder communications is a key limitation which arises not so much due to weaknesses within each of the stakeholder groups themselves. Rather, it is the disconnection between some stakeholders and limited recognition of the role that the community and the tourism industry play in the context of fire planning, emergency management and recovery by authorities that contribute to future vulnerability. In particular, there has been little recognition of visitors and future visitors as a stakeholder group who are essential to future social and economic resilience.

Stakeholder theory is underpinned by two main arguments. First, is the practical claim that diverse stakeholder interests and perspectives need to be identified, understood and managed in the interests of sustainability of a given context such as an organisation or, in this case, a small destination tourism economy (Byrd, 2007). This practical imperative arises given that stakeholders have a legitimate interest in aspects of the functioning of the tourism system, giving the stakeholder power to affect economic performance and/or has a stake in performance (Sautter & Leisen, 1999, p. 313). Thus, the system or organisation is made vulnerable where stakeholder relationships are fragmented and/or conflictual due to stakeholder perspectives being overlooked, undervalued or misunderstood. The second argument is normative. That is that all stakeholders, with legitimate interests in an organisation or organisational system, deserve consideration for, '... its own sake and not merely because of its ability to further the interests of some other group, such as the shareowner.' (Donaldson & Preston, 1995, p. 67).

The V/R analysis of Harrietville following the 2013 bushfire is revealing in relation to the practical importance of engagement with, and communication between, all tourism industry stakeholders as partners in maximising the effectiveness of emergency planning, response and recovery. As discussed, the key vulnerabilities brought to light occurred due to misunderstandings, gaps in communication and a lack of appreciation of each stakeholder's role in preserving nature-based assets and the tourism values of the area. While there is little disagreement about the need for emergency management to be completely focussed on the safety of lives and property, better planning, in consultation with all stakeholders, could have minimised reputational damage to the town and some of the stressors and costs experienced by community members and business operators. Similarly, tourism operators could better appreciate the limitations of government and take greater responsibility for bushfire preparedness. Further, better planning and communications with visitors and future visitors may have minimised the reputational damage to the town and decreased the risks faced by particular groups of visitors.

As identified by Sautter and Leisen (1999, p. 315), stakeholder theory requires the identification of stakeholders and their interests in the tourism system. In the case of Harrietville, key stakeholders can be usefully summarised within five key stakeholder groups, each with a particular interest in the tourism economy. As illustrated in Figure 1 below, key stakeholders include first, the community which comprises Harrietville residents, community associations and residents employed within the tourism industry. The tourism industry is seen to comprise tourism business operators, suppliers and tourism planning authorities. Visitors and future visitor are a key stakeholder given that visitor spending drives the tourism economy. Government and land management authorities are responsible for town governance and the management and preservation of the nature-based assets of the area. Emergency planning, response and recovery in the state of Victoria is the responsibility of multiple agencies, both government and volunteer with the primary interest of protecting lives and property. As such, Figure 1 illustrates Emergency Services as being closely, yet indirectly connected to the tourism economy. While emergency services have a direct impact on tourism assets and people engaged in tourism, the tourism industry, per se, is not the central focus. Figure 1 also illustrates the communications flow in relation to emergency planning, response and recovery as a means to illustrate the potential blockages in information flow in relation to tourism management.

As shown, communications between stakeholders prior to the events of 2013 in Harrietville lacked integration. The decisions and actions of emergency management services are planned and undertaken by government, yet have been made largely in isolation from the broader community and tourism industry. This is despite the impacts that all emergency management actions have on the tourism economy. Similarly, there is little communications with visitors who are largely reliant on information provided by tourism operators and, to a lesser extent, community members. This lack of communication means that many visitors are vulnerable to being exposed unnecessarily to bushfire risk and are less likely to be able to interpret the accuracy and reliability of messages generated by media. Most important is that each stakeholder group previously operated separately with little opportunity for collaboration.

As discussed, government agencies have since supported the development of mechanisms to improve stakeholder engagement and joint participation in planning. The major initiative is the establishment of the Harrietville Community Forum which is led by an

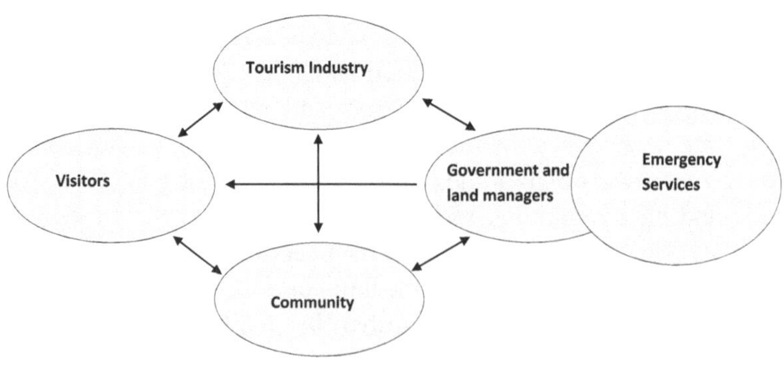

Figure 1. Tourism stakeholders and communication flows.

elected committee made up of community members and tourism operators. This is the mechanism for community participation in government planning and policy development and to build community resilience. The evolution of the forum was widely identified as a success. As one forum member commented,

> Out of this fire came the Harrietville community forum and out of that a decision was made that we needed to repair and build the relationships with agencies. The Harrietville Community Forum is now an incorporated body, which is important because people in the town then become members and we work closely with the other agencies. That's probably the biggest single advance that we've made is with the cooperation, collaboration and working alongside of all of the other agencies that work here – you know, firefighters, emergency services, Parks & Wildlife, State Forest, Victoria Roads, council – all of those people.

Overall, the need for improved stakeholder engagement has been widely recognised as central to building the resilience of the town, and of sustaining the long term viability of the nature-based tourism economy.

Conclusion

The effects and aftermath of 'the Harrietville fire' in 2013 provide an important case study in terms of understanding vulnerability and resilience of the tourism economy to a major shock to the system. Through the lens of the DSF, a number of factors contributed to vulnerability. Fragmented stakeholder communications and integration in planning was a key factor that impeded recovery and contributed to the impacts of the event.

The need for improvements in this regard has been widely recognised and significant actions have been undertaken to strengthen engagement by key stakeholders. The tourism industry, community members, governance authorities and emergency management agencies now have a forum through which to engage in collaborative planning and policy development in relation to emergency planning, response and recovery. While this forum is a 'work in progress', it is a measure that has been deemed to be highly successful and has given cause for optimism by all stakeholders. The extent to which the interests of visitors are represented or understood, however, remains tenuous. While visitors can be represented by tourism industry representatives, effective understanding of, and collaboration with, visitors continues to represent a gap in stakeholder engagement. How this is most effectively approached requires further examination and research.

In addition to the specific findings, this case study has wider implications for nature-based destinations elsewhere, as it offers a model for replication in other communities and contexts. The study also highlights the great value of 'bottom up' approaches to planning and, in the interests of sustainability, the crucial importance of connecting with the 'local' as deeply as possible.

This is in contrast to the traditional and widely used 'top down' approach, of assessing risk by external, quantitative means and imposing risk abatement strategies from above on communities.

Disclosure statement

No potential conflict of interest was reported by the authors.

References

Alpine Shire Council. (2011). *Alpine Shire business survey 2011*. Bright, Victoria: Author. Retrieved from http://www.alpineshire.vic.gov.au/files/Business/Business_Survey_FINAL.pdf

Alpine Shire Council. (2013). *Market profile summer 2013: Bright Myrlteford mount beauty Harrietville*. Bright, Victoria: Author. Retrieved from http://www.alpineshire.vic.gov.au/files/2013Summer_Statistics.pdf

Bird, D., Gísladóttir, G., & Dominey-Howes, D. (2011). Different communities, different perspectives: Issues affecting residents' response to a volcanic eruption in southern Iceland. *Bulletin of Volcanology*, *73*(9), 1209–1227. doi:10.1007/s00445-011-0464-1

Bourne, L. (2008). Project relationship management and the stakeholder circle™. *International Journal of Managing Projects in Business*, *1*(1), 125–130.

Bramwell, B., & Sharman, A. (1999). Collaboration in local tourism policymaking. *Annals of Tourism Research*, *26*(2), 392–415. doi:http://doi.org/10.1016/S0160-7383(98)00105-4

Byrd, E. T. (2007). Stakeholders in sustainable tourism development and their roles: Applying stakeholder theory to sustainable tourism development. *Tourism Review*, *62*(2), 6–13. doi:10.1108/16605370780000309

Calgaro, E. (2010). Building resilient tourism destination futures in a world of uncertainty: assessing destination vulnerability in Khao Lak, Patong and Phi Phi Don, Thailand to the 2004 Tsunami. Sydney: (Pacific Tourism – Climate Adaptation Project No. 76). Department of Environment and Geography, Macquarie University.

Calgaro, E., Dominey-Howes, D., & Lloyd, K. (2014). Application of the destination sustainability framework to explore the drivers of vulnerability and resilience in Thailand following the 2004 Indian Ocean Tsunami. *Journal of Sustainable Tourism*, *22*(3), 361–383. doi:10.1080/09669582.2013.826231

Calgaro, E., Lloyd, K., & Dominey-Howes, D. (2014). From vulnerability to transformation: A framework for assessing the vulnerability and resilience of tourism destinations. *Journal of Sustainable Tourism*, *22*(3), 341–360. doi:10.1080/09669582.2013.826229

Carpenter, S., Walker, B., Anderies, J. M., & Abel, N. (2001). From metaphor to measurement: Resilience of what to what? *Ecosystems*, *4*(8), 765–781.

CFA. (2014). Harrietville community information guide – Bushfire. In Country Fire Authority (Ed.). Retrieved from http://cfaonline.cfa.vic.gov.au/mycfa/Show?pageId=publicDisplayDoc&fname=2014/CIG-W-HUM-Harrietville-2_00_3017.pdf

Donaldson, T., & Preston, L. (1995). The stakeholder theory of the corporation: Concepts, evidence, and implications. *Academy of Management*, *20*(1), 65–91.

Fletcher, R. (2009). Ecotourism discourse: Challenging the stakeholders theory. *Journal of Ecotourism*, *8*(3), 269–285. doi:10.1080/14724040902767245

Freeman, R. (1984). *Strategic management: A stakeholder approach*. Boston, MA: Pitman.

Gaillard, J. C., & Mercer, J. (2013). From knowledge to action: Bridging gaps in disaster risk reduction. *Progress in Human Geography*, *37*(1), 93–114. doi:10.1177/0309132512446717

Hallowes, M. (2013). *Harrietville fire 2013: Report to the minister for police and emergency services*. Melbourne, Victoria: Office of the Emergency Services Commissioner, State Government of Victoria Department of Justice.

Hughes, L., & Steffen, W. (2013). *Be prepared: Climate change and the Australian bushfire threat*. Potts Point, NSW: Climate Council of Australia Ltd.

IPCC. (2007). *Climate change 2007: Impacts, adaptation and vulnerability, contribution of working group II to the fourth assessment report of the IPCC*. New York, NY: Intergovernmental Panel on Climate Change (IPCC). Retrieved from https://www.ipcc.ch/publications_and_data/ar4/wg2/en/contents.html

IPPC. (2012). Summary for policymakers. In *Managing the risks of extreme events and disasters to advance climate change adaptation. A special report of working groups I and II of the intergovernmental panel on climate change*, Cambridge, UK: Cambridge University Press. Retrieved from https://www.ipcc.ch/pdf/special-reports/srex/SREX_FD_SPM_final.pdf

ISO Guide 73. (2009). *Risk management - Vocabulary*. International Organization for Standardization. Retreived from http://www.iso.org/iso/catalogue_detail?csnumber=44651
Jiang, M., & DeLacy, T. (2014). A climate change adaptation framework for Pacific Island tourism. In T. DeLacy, M. Jiang, G. Lipman, & S. Vorster (Eds.), *Green growth and travelism: Concept, policy and practice for sustainable tourism* (pp. 225–240). London, UK: Routledge.
Jopp, R., De Lacy, T., & Mair, J. (2010). Developing a framework for regional destination adaptation to climate change. *Current Issues in Tourism*, *13*(6), 591–605. doi:10.1080/13683501003653379
Jopp, R., De Lacy, T., Mair, J., & Fluker, M. (2013). Using a regional tourism adaptation framework to determine climate change adaptation options for Victoria's Surf Coast. *Asia Pacific Journal of Tourism Research*, *18*(1-2), 144–164. doi:10.1080/10941665.2012.688515
Kasperson, R. E., Turner, B. L., Schiller, A., & Hsieh, W.-H. (2001). *Research and assessment systems for sustainability: Framework for vulnerability*. Paper presented at the the meeting of AIACC Project Development Workshop: Climate Change Vulnerability and Adaptation Trieste, Italy.
Klint, L. (2012). *"Buoyancy – bifo and afta": A climate change vulnerability/resilience framework for tourism – The case study of Vanuatu dive tourism. PhD Thesis*. Victoria University, Melbourne, Australia.
Klint, L., Jiang, M., Law, A., De Lacy, T., Filep, S., Calgaro, E., ... Harrison, D. (2012). Dive tourism in Luganville, Vanuatu: Shocks, stressors, and vulnerability to climate change. *Tourism in Marine Environments*, *8*(1/2), 91–109. doi:http://doi.org/10.3727/154427312X13262430524225
Mercer, J., Gaillard, J. C., Crowley, K., Shannon, R., Alexander, B., Day, S., & Becker, J. (2012). Culture and disaster risk reduction: Lessons and opportunities. *Environmental Hazards*, *11*(2), 74–95. doi:10.1080/17477891.2011.609876
Mercer, J., Kelman, I., Taranis, L., & Suchet-Pearson, S. (2010). Framework for integrating indigenous and scientific knowledge for disaster risk reduction. *Disasters*, *34*(1), 214–239. doi:10.1111/j.1467-7717.2009.01126.x
Moreno, A., & Becken, S. (2009). A climate change vulnerability assessment methodology for coastal tourism. *Journal of Sustainable Tourism*, *17*(4), 473–488. doi:10.1080/09669580802651681
North East Victoria Tourism Board. (2012). *North East Victoria tourism board: Strategic plan and implementation actions 2012–2014*. Beechworth, VIC: Author.
Palttala, P., Boano, C., Lund, R., & Vos, M. (2012). Communication gaps in disaster management: Perceptions by experts from governmental and non-governmental organizations. *Journal of Contingencies and Crisis Management*, *20*(1), 2–12. doi:10.1111/j.1468-5973.2011.00656.x
Patton, M. Q. (2002). *Qualitative research & evaluation methods* (3rd ed.). Thousand Oaks, CA: Sage Publications Inc.
Polsky, C., Neff, R., & Yarnal, B. (2007). Building comparable global change vulnerability assessments: The vulnerability scoping diagram. *Global Environmental Change*, *17*(3–4), 472–485. doi:http://doi.org/10.1016/j.gloenvcha.2007.01.005
Pyke, J., Jiang, M., de Lacy, T., Whitelaw, P., & Jones, R. (2015). *Assessing the economic value and vulnerability of nature based tourism in the ovens and alpine area of NE Victoria*. Melbourne, Australia: Victoria University, Melbourne and the Cooperative Research Centre for Bushfires and Natural Hazards.
Sanders, D., Laing, J., & Houghton, M. (2008). *Impact of bushfires on tourism and visitation in alpine national parks*. Gold Coast, Australia: CRC for Sustainable Tourism.
Sautter, E. T., & Leisen, B. (1999). Managing stakeholders a tourism planning model. *Annals of Tourism Research*, *26*(2), 312–328. doi:http://doi.org/10.1016/S0160-7383(98)00097-8
Scott, D., Amelung, B., Becken, S., Ceron, J. P., Dubois, G., Gössling, S., ... Simpson, M. (2008). *Climate change and tourism: Responding to global challenges*. Madrid, Spain: UNWTO and UNEP.
Tourism Research Australia. (2014). *National visitor survey*. Retrieved from http://www.tra.gov.au/aboutus/national-visitor-survey.html
Waligo, V. M., Clarke, J., & Hawkins, R. (2013). Implementing sustainable tourism: A multi-stakeholder involvement management framework. *Tourism Management*, *36*(0), 342–353. doi:http://doi.org/10.1016/j.tourman.2012.10.008

World heritage listing as a catalyst for collaboration: can Mount Fuji's trail signs point the way for Japan's multi-purpose national parks?

Thomas Jones, Sue Beeton and Malcolm Cooper

ABSTRACT

Iconic Mount Fuji has symbolized the fragmented state of Japan's 'multi-purpose' national park administration. Although the Ministry of Environment (MOE) is the legal administrator, the national parkland is predominantly owned by the Forestry Agency, with many services provided by local governments. This complex combination can result in stakeholders pitted against each other, but listing as a World Heritage Site (WHS) offers an external incentive toward holistic management. This paper reviews Fuji's 2013 WHS inscription process, analyzing official documents and a series of stakeholder interviews to substantiate claims that it helped catalyze stakeholder collaboration, as demonstrated by a new system of trail signs. A cross-cutting council formed in 2009 issued a set of guidelines aimed at standardization of place names, and the removal of unnecessary or low-quality signs. Subsequently, a simplified system of color-coded signs was introduced along the four climbing trails, epitomizing the new spirit of cooperation. However, more research is needed to verify if the new partnership can extend the collaboration from trail signs to incorporate broader provision of visitor services, and now that the 'carrot' of inscription has been achieved, the multi-stakeholder momentum may be difficult to maintain.

Introduction

Given Curry's (1994) assertion that 'access is synonymous with conflict,' the continued growth of tourism to protected areas increases the likelihood of rifts occurring. National park administrators are therefore increasingly required to manage relationships and incentivize partners to catalyze stakeholder collaboration. At an inter-institutional level, participative decision-making is the key to coordinating conservation and development (Beeton, 2006; Hammer, Mose, Siegrist, & Weixlbaumer, 2007; Heritier, 2010). However, a related but under-explored alternative is the external 'carrot' of inscription on UNESCO's list of World Heritage Sites (WHS) which could help generate multi-stakeholder momentum around a common vision.

This paper investigates a set of indirect impacts linked to Mt. Fuji's belated inscription as a WHS in 2013. A series of stakeholder interviews is used to substantiate claims that the

WHS inscription process helped usher in a new era of improved stakeholder collaboration, epitomized by a new system of trail signs. Fuji is Japan's highest mountain at 3776 m and was designated within the Fuji-Hakone-Izu National Park in 1936. Fuji was also among the first candidate sites on the tentative list after Japan ratified the WH Convention in 1992, but the initial movement to apply for *natural* WHS status was aborted and the inscription application re-booted to emphasize Fuji's *cultural* heritage aspects.

Building on previous research that quantifies WHS inscription impacts, this paper therefore revisits the listing process' potential to catalyze stakeholder collaboration. To achieve this aim, a mixed-method approach is structured around a single case study that follows Yin (2009, p. 8)'s definition of an empirical inquiry that 'investigates a contemporary phenomenon in depth and within its real-life context.' Fuji's selection as the case study site is justified by the insights it offers into the intersection between UNESCO's WHS program and Japanese national park management. The methodology employed official reports and meeting minutes, supported by interviews with key stakeholders who were closely involved with the WHS inscription process. A pilot interview was conducted in November 2013 and the remainder carried out from April 2014 to August 2015. All 16 semi-structured interviews lasted from 30 to 90 min. 11 were conducted at the respondent's workplace, four in the field, and one over the telephone. Haukeland (2011) noted that this kind of mixed-method qualitative research remains a relatively uncommon approach despite facilitating a 'rich and detailed examination of stakeholder relationships' (Wilson, Nielsen, & Buultjens, 2009). The interviews comprised a cross-cutting selection of stakeholders to the extent that additional observation would not yield any substantially different information. Findings were further contextualized by direct observation and stakeholder documents, predominantly via Japanese language sources. The approach thus follows the principle of triangulation through multiple sources of empirical evidence (Yin, 2009). This study adds to the increasing body of work on stakeholder collaboration in protected areas by tackling a gap in the literature that investigates the importance of UNESCO's WHS program as an external conflict-resolution mechanism which can help catalyze the creation of heritage management partnerships (Jones, 2012).

Stakeholder conflict and co-existence

Stakeholder conflict in protected areas

As tourism's role in revitalizing rural areas has become increasingly important, protected areas such as national parks face growing pressure to utilize the market mechanisms involved in this industry as a tool for conservation (Pascual & Perrings, 2007), and to demonstrate their effective management to justify budgets (Eagles, 2002). Increasingly professional visitor services are expected of parks and heritage sites even as management flashpoints such as trails and toilets are strained by the additional footfalls. Behind the scenes, the growing emphasis on providing 'quality' tourism services has made it increasingly important to manage relationships (Sautter & Leisen, 1999), and to explore partnerships (Kuvan & Akan, 2012). In order to do so, the role of stakeholders (Fennell & Weaver, 2005) needs to be clearly identified as a step on the way to improving collaboration – defined as 'a process of joint decision-making among autonomous, key stakeholders of an inter-organizational, community tourism domain to resolve planning problems of the domain and/or to manage issues related to the planning and development of the domain' (Jamal & Getz, 1995).

It is this joint decision-making that is portrayed as the key to coordinating conservation and development (Hammer et al., 2007). Yet, for many of the world's protected areas, Curry's (1994) assertion that 'access is synonymous with conflict' remains more relevant than ever. And although conflict is not confined to tourism, the rapid socio-economic changes associated with the tourism sector are conducive to fragmentation of stakeholders, as in the example of recreation and agriculture in the UK (MacMillan & Phillip, 2008). Sources of conflict can also be traced back to incompatible institutional objectives, such as in the national parks of England and Wales, where the access-conservation dilemma repeatedly pits tourism development agencies and DMOs against conservationists. Heightened awareness of the struggles on both sides has yet to reveal a long-term resolution to this impasse, and if anything, 'these two purposes have increasingly come into conflict' (Sharpley, 2009, p. 161). Beyond tourism, examples of conflict – and co-existence – frequently derive from multi-stakeholder scenarios as in Scotland, where upland landscapes have conventionally played host to multiple land uses including sheep and cattle farming, together with forestry and hunting of grouse or deer on sporting estates (Morgan-Davies & Waterhouse, 2010).

Given such complicated combinations of actors, the 'round-table' approach of multi-stakeholder participation in decision-making remains an unproven but universally appealing option. However, because communities have distinct cultural characteristics, a separate set of consensus-building tools is required for each scenario (Axelrod, 1997; Mitchell-Banks, 1997). In the context of the US Forest Service land, for example, Blahna and Yonts-Shepard (1989) identified numerous barriers to providing effective public participation, including '(1) the complex, technical planning process adopted by the agency; (2) a lack of agency guidance on conducting interactive public participation; (3) a desire to avoid controversy; and (4) internal power struggles that affected the planning process.' Stewart and Sinclair (2007) argue that participants themselves often criticize the round-table approach as dissatisfying, time-consuming, costly, and inefficient. In short, the process of bringing a cross-section of stakeholders into the organic decision-making fold has certain shortcomings which ensure that the potential of external 'carrots' to catalyze stakeholder collaboration cannot be ignored. One of the most prestigious is the inscription of a destination as a UNESCO World Heritage Site (WHS). Described by Buckley (2004) as a 'top brand,' the perceived quality and distinctiveness of WHS status is thought to confer a competitive advantage on listed destinations that could motivate stakeholders and generate collaborative momentum in the run-up to inscription (Ryan & Silvanto, 2009).

However, despite such incentives and the kudos of listing, WHS status provides no extra budget or formal jurisdictional rights, so previous research is divided over whether it is in fact advantageous for destinations. To that end, an extensive literature review by Poria, Reichel, and Cohen (2011) found 'no clear evidence that the title WHS has strong net brand equity in terms of increase in the number of visitors.' Similarly, empirical results from Macau found no significant effect of WHS listing on tourism promotion (Huang, Tsaur, & Yang, 2012), while in Barcelona, Palau-Saumell, Forgas-Coll, Sánchez-García, and Prats-Planagumà (2013) also concluded that 'the WHS label … is not fundamental for attracting millions of visitors.' The benefits of listing are even more contentious in protected areas where fragmented management arrangements may be further strained as increasing visitor numbers result in heightened environmental or social impacts. Inscription does entail the creation of a management plan to counter such environmental impacts as waste and trail erosion, or social ones such as congestion and commercialization (Jimura, 2011). However the uncertain economic benefits and negative impacts, often exacerbated by WHS inscription, should be set against the potential for improved stakeholder collaboration. This paper next examines the case of Japan's national parks, and Fuji-Hakone-Izu in particular, to develop the argument that the WHS listing process could offer a powerful external incentive toward more holistic management.

A case for co-existence in Japan's national parks

The presence of multiple land owners ensures that stakeholder collaboration is of vital importance to Japan's national parks, which are administered by the Ministry of Environment (MOE) under the Nature Parks Act.[1] The MOE uses zoning and regulatory tools to prioritize favored management objectives such as conservation and public access over other forms of land use such as farming and forestry. However, the 'multi-purpose' system faces multiple impediments to holistic management, including institutional overlap and difficulties in extrapolating a common vision from the tangled web of multiple stakeholders. Hiwasaki (2005) notes that 'in order to reach the system's potential, the park authority must act as coordinator of stakeholders.' This echoes the Mahanty and Russell (2011) argument that 'conservation professionals need to build their capacity as facilitators and negotiators, paying greater attention to how stakeholder groups form and function, their links to wider arenas, and the aims and positions of groups and individuals.' However, the MOE's capacity for such coordination is complicated by the highly complex stakeholder networks created by national parks' 'multi-purpose' agenda and mixed land ownership. Provision of visitor services such as trails and toilets has subsequently depended on *de facto* partnerships of public and private stakeholders to compensate for the lack of a 'one-stop shop' equivalent to the U.S.A.'s National Park Service (Murakushi, 2005). Although the first reference to co-operation appeared in a MOE management plan in 1979, and collaborative management took hold in 1996, it was not until the 2008 amendments to the Nature Parks Act were implemented the following year that a 'partnership' ethos was formalized as a core management concept (Watanabe, Sasaki, Shinohe, & Shimomura, 2012). The collaborative management concept has subsequently assumed increasing importance in line with national and global calls to resolve fragmented governance (Tanaka, 2014), and establish a platform for regular exchange of information and multi-stakeholder dialogue (Aikoh, 2014).

Key stakeholders in Japan's national parks

The MOE's capacity to achieve coordination of national park stakeholders remains reliant on their interaction with a range of related actors, including other central government institutions such as the Forestry Agency, a significant owner of national parkland, but one whose institutional objectives – including commercial logging – can seem incompatible with those of the MOE (Natori, 1997). Local government also has an important stake, often in the form of historical land ownership or use rights upheld by prefectures or municipalities. Meanwhile private sector involvements ranging from transport and accommodation conglomerates to real estate agents, marketing and travel agencies control considerable swathes of the economic impact attributed to nature-based tourism. Given the complicated list of stakeholders involved in just one council (see Appendix 1) a comprehensive inventory of all actors is beyond the parameters of this paper. The analysis thus focuses on two integral horizontal and vertical tiers (Cawley & Gillmor, 2008), examining two principal central government actors – the MOE and MAFF – before discussing the two Prefectural governments of Shizuoka and Yamanashi within whose borders the Fuji district of the Fuji-Hakone-Izu National Park is located.

The ministry of environment (MOE)

Japan's national parks are administered by the Ministry of Environment (MOE) under the Nature Parks Act (Katō, 2008). In lieu of land ownership, the MOE uses zoning and regulation to achieve their institutional objectives, with a focus on conservation and public access. By 2004, 260 Rangers and 80 Active Rangers were employed across 32 parks, so a volunteer workforce of Nature Park Leaders, Park Volunteers and Green Workers is also important in supporting the eradication of invasive species programs or rehabilitating degraded ecosystems. There have been recent attempts to introduce a more pro-active approach to visitor management, for example by bolstering on-site presence with more Active Rangers (Katō, 2008). However, the MOE's logistical remit still consists chiefly of administrative tasks such as formulating park plans, issuing permits and approving planning permission applications.

Each national park has an MOE management plan, updated periodically and divided into a zoning system that sub-classifies the parkland into three conservation bands, ranging from Special Protection Zones, to Special Zones (sub-divided into Class I, Class II and Class III) and Ordinary Zones. These are equivalent to core and buffer zones (Ito, 1996), with Special Protection Zones at the top of the hierarchy subject to the strictest regulations (for example, MOE permission is needed for any kind of cutting or collecting wood, plants, or rocks). In Class I-III Special Zones, commercial activities such as construction, forestry, and even vehicular access (by snow mobiles and so on) are still regulated by the MOE. However, in the Ordinary Zones, development is rarely restrained by MOE's regulations, although mining, advertising hoardings and large-scale construction projects do require MOE notification. Moreover, holistic implementation of the zoning system faces impediments due to the MOE's lack of land ownership,[2] funding and human resources (Imura, 2005). Hiwasaki (2005) argues that to operationalize multiple-use park management, the MOE 'must act as coordinator of stakeholders.' However, coordinating conservation-based objectives can be complicated by the lack of alignment with the goals of other central government agencies such as the MAFF.

The ministry of agriculture, forestry and fisheries (MAFF)

67% of the Japanese archipelago is forested, and another significant national park stakeholder is the Forestry Agency housed within the Ministry of Agriculture, Forestry and Fisheries (MAFF). The MAFF controls 62% of all national parkland stemming from their jurisdiction over national forest, which in turn contain 95% of Japan's UNESCO natural heritage sites,[3] and over 75% of the national parks' Special Protected Zones (Forestry Agency, 2006). This reflects the historical prevalence of 'secondary nature' within the national parks, many of whose landscapes were shaped by traditional uses such as farming and forestry. However, the MAFF's managerial objectives have rarely been contiguous with those of the MOE, conventionally favoring logging over conservation or tourism (Katō, 2008, p. 53). This conflict of interests undermines the MOE's capacity for coordinated park planning, as evidenced by research in Hokkaido which used GIS to overlay a national forest management plan with that of a national park. The results showed that around half of the national park boundaries were indistinguishable from those of the national forest, suggesting that the park border – which should have been delineated to safeguard appropriate conservation sites – was in reality zoned around *a priori* interests such as timber extraction that is inconsistent with the MOE's objectives (Aikoh & Tomidokoro, 2010).

More noticeably, the Forestry Agency's historical right to log areas of national forest within the national parks has also resulted in periodic conflicts, as with the Shiretoko logging controversy which played out from 1981 to 1987 after the announcement of a plan to clear-cut areas of Shiretoko National Park (Natori, 1997). The Forestry Agency's purported aim of sustainable timber production and regeneration proved anathema to a conservation alliance that was initiated by the Shiretoko National Trust movement, but ultimately centered on the MOE. Here was a rare example of the MOE departing from its conflict-averse stance to argue strongly in favor of ecosystem protection, citing the unique nature of one of Japan's last-surviving wilderness areas (Mitsuda & Geisler, 1992). The Shiretoko incident helped accelerate the diversification of Japanese forest policy away from monoculture plantations, with increasing reference made thereafter by the Forestry Agency to 'alternative' goals including biodiversity and recreation. However, the change in tack owed as much to the ongoing decline in the forestry industry[4] as to a fundamental re-calibration of MAFF policy objectives, and improvement in inter-institutional collaboration therefore remained largely symbolic (Tanaka, 2014).

The role of other central and local government agencies

National park planning has also been shaped by other government agencies. The Ministry of Land, Infrastructure and Transport (MLIT)[5] has jurisdiction over large roads and rivers, as well as disaster management infrastructure such as sand control dams and rubble retention fields. While MAFF is the main landowner, MLIT at one point controlled over 80% of the central government's public works budget[6] (Feldhoff, 2002). Aside from fiscal muscle, MLIT and affiliates' proximity to civil engineering firms also ensured close ties to private corporations (Knight, 2010) fueling a 'construction state' to which Mason (1999) attributed '[m]any of Japan's domestic environmental problems.'

Aside from the tightrope between conservation and development, another dimension to the park administration balancing act is shaped by the struggle between local government and the state. Local government actors owned 12.4% of Japan's total parkland in 2011 and also play a vital role in the everyday running of the parks, contributing human and fiscal resources to help maintain a diverse range of visitor amenities including car parks, trails and toilets (Tanaka, 2014). Unlike the vertical networks of central government stakeholders that dominate national park administration, local governments share contiguous geographic borders which could result in a more compact and closely-aligned decision-making unit with heightened opportunities to interact, for example via regional disaster management drills (Jones, 2015). Yet many local authorities in Japan are burdened by debt or even bankruptcy, as in the 2006 case of Yubari in Hokkaido (Seaton, 2010). A climate of fiscal austerity has prevailed, and although the 2003 trinity reform package nominally shifted the responsibility of national park administration towards a more centralized mode of governance, local governments were still left with considerable financial responsibilities to the extent that several parks – including Fuji – have had to contemplate the introduction of user fees (Katō, 2008).

The role of non- government actors

Private sector also plays a significant role in national park management (Beeton, 2002) and Fuji's massive visitor footfall ensures that tourism is the dominant economic sector (Chakraborty & Jones, 2018). 26% of all national parkland is privately owned, and tourism services such as transport, accommodation, and catering are provided by private operators. Historically, such stakeholders have held such sway over the park that designation could perversely 'prove detrimental to the conservation of the area, owing to such factors as the development of tourist facilities, road construction, vehicle pollution and over-use' (Ishikawa, 2001; Natori, 1997; cited in Knight, 2010). Although less influential than during the heyday of Japan's 'economic miracle' in the 1960s, tourism remains a significant stakeholder sector.

Non-government organizations represent another group with growing sway in Japan's national parks. Opportunities for participation have been increasing, particularly since the 1995 Hanshin earthquake helped kick-start the re-emergence of civil society (Evans, 2002). This reflected the global trends epitomized by the creation of the UN's Agenda 21 to promote the participation of civil society in environmental decision-making (Schreurs, 2005). However, park administration still offers 'no formal opportunities that guarantee the incorporation of voices in decision-making' (Hiwasaki, 2006), relying instead on informal partnerships or reciprocal arrangements. Consequently Japanese environmental NGOs were comparatively less integrated into pluralistic networks and therefore play a less substantial role in the policy-making process, exacerbated by the relatively paucity of members, staff and financial resources (Foljanty-Jost, 2005). Here our case study site offers an encouraging example, with the Fujisan Club established in 1998. Using a mixture of individual ($n = 1{,}200$) and corporate members ($n = 77$), the NGO has mobilized volunteers on both sides of Fuji to participate in clean-up activities, environmental education programs and special campaigns to remove invasive species.

Case study of Mount Fuji

Mt. Fuji, Japan's highest peak at 3776 m, is located some 120 km southwest of Tokyo between the prefectures of Yamanashi and Shizuoka. Fuji's iconic shape derives from a cone-shaped stratovolcano whose slopes rise continuously from the Pacific Ocean to the summit. Revered from afar since ancient times, a climbing culture emerged as worshippers sought out sacred sites on and around the mountain, and in the fourteenth century Shugendo practitioners established a trail that led pilgrims to the summit (Polidor, 2007). Thereafter the peak became the goal for guided groups of pilgrims, whose numbers were estimated at over ten thousand in 1877 (Murakushi, 2005). During the post-war period of extended economic growth, climber numbers increased rapidly. Rising living standards brought both increased leisure time and improved access infrastructure, including the construction of many toll roads within national parks (Oyadomari, 1989). At Fuji, the Subaru line in Yamanashi was completed in 1964, providing direct access to the fifth station (approximately 2300 m ASL) and dramatically reducing the round-trip climbing time.

Today, the three-month summer season draws over 300,000 climbers (Jones, Yang, & Yamamoto, 2018). Together with the three million annual visitors to the 5th station trailhead, this has resulted in a cluster of management issues including congested trails, and toilets and trash (Sayama & Nishida, 2001). Fuji's iconic attractiveness and proximity to the urban hubs – in particular, Kanto (the greater Tokyo metropolitan area) – ensures visitation that accounts for a third of all national park visits in the country. Yet holistic management strategies are elusive due to fragmented stakeholders. Within the Fuji region of the national park, 41% of the land belongs to prefectural and 38% to private owners. The MOE presence in terms of land ownership is minor, amounting to two branch offices, toilets on the summit and descending trail, and the summit loop trail. This pales into insignificance compared to the 21% national forest that falls under MAFF jurisdiction. 60% of this consists of plantation forests, with half comprised of a single species, Hinoki Cypress (*Chamaecyparis obtuse*) that was once a linchpin of the forestry industry. Other central government agencies include the Ministry of Education (MEXT, whose Agency of Cultural Affairs facilitated the application process for cultural WHS); the Ministry of Interior, which owns the land on which some UNESCO components are located; and the Self-Defence Forces which took over management of the East Flank military maneuver areas from the U.S. in 1958 (Bernstein, 2013). The base accounts for the missing chunk on the North-East flank (Figure 1). The designated national parkland also overlaps with the administrative territory of 15 municipal, and two prefectural governments.

It was these two prefectures – Shizuoka and Yamanashi – that would play the pivotal role in Fuji's WHS listing, yet the successful application papered over stark divides in their geographical, economic and socio-political circumstances. Shizuoka Prefecture covers an area (7,780 km^2), almost double the size of Yamanashi Prefecture (4,465 km^2) resulting in a distinct advantage in terms of population (3,700,000) and tax revenue ($4 billion) compared to Yamanashi (840,000; $800 million according to Noguchi, 2014, p. 59).Yamanashi's remote, mountainous villages were historically less affluent than the southern fishing and farming communities in Shizuoka, which also possessed some well-known hot spring resorts along the main Tokai trade route (Totman, 2007). The geological composition of the northern flank in Yamanashi created water scarcity that impeded the spread of rice farming (Kureha, Kikuchi, Sano, & Yamamoto, 2015). Instead, Fuji's Lake District had to pin its hopes on tourism from an early stage of development, resulting in various land policies such as the 1916 Hokuroku Tourism Development Plan (Tsuchiya, 1981). In the post-war period, Shizuoka's location along the Pacific coast placed it within the chain of large-scale, heavy industry clusters that were the engine of Japan's extended period of economic growth. The southern row of steel and petrochemical factories, together with electronics and precision goods manufacturers, stood in contrast to

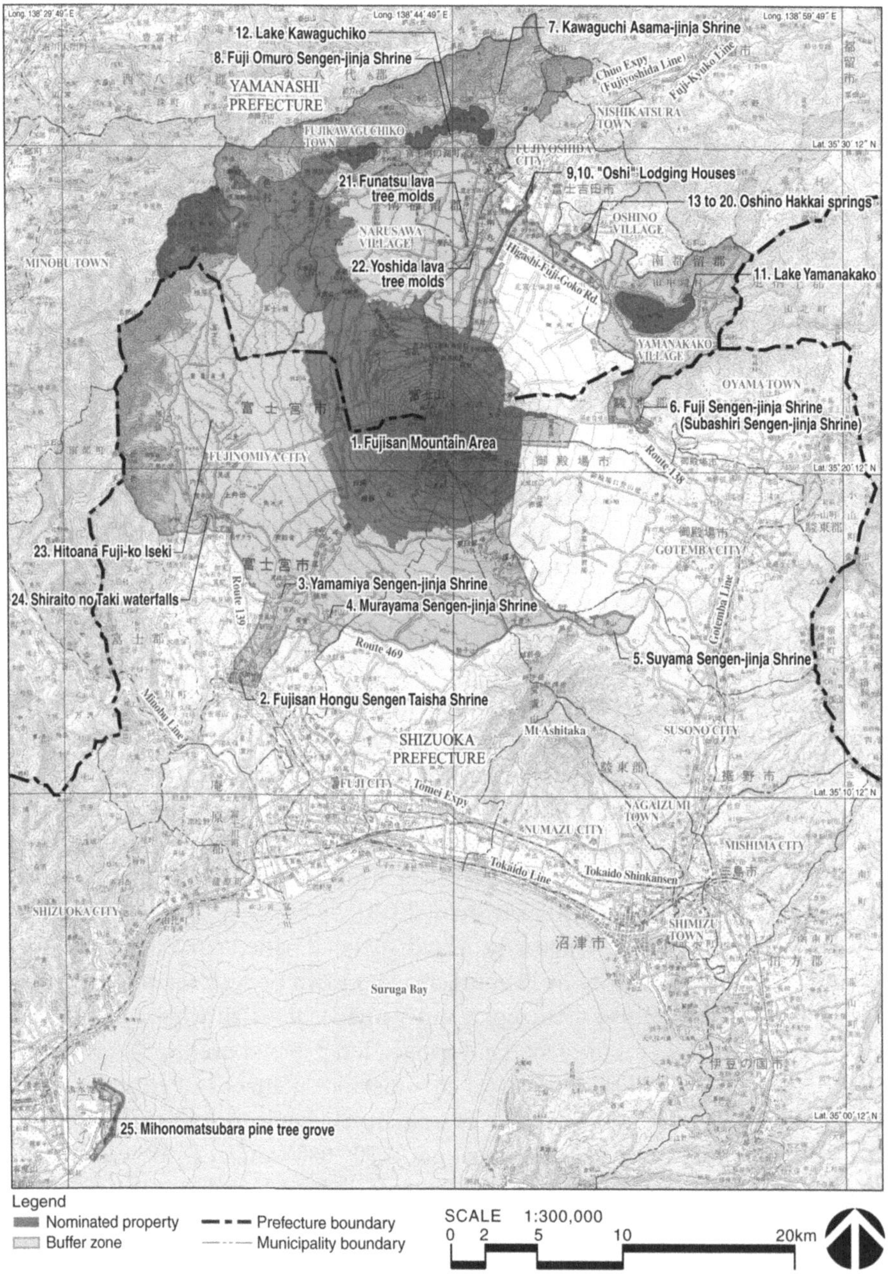

Figure 1. Map of the components of Fujisan inscribed on the World Heritage List (adapted from the nomination dossier).

Yamanashi, where the economy was less diversified and more reliant on seasonal tourism. The fact that the *shinkansen* bullet train line runs south of Fuji through Shizuoka is portrayed as a further reminder of the gap between the prefectures' perceived prestige.

These distinct socio-economic situations form the backdrop for the two prefectures' respective standpoints in the perennial park management balancing act. One of the longest-running examples of stakeholder conflict at Fuji is the summit ownership rights court case that ran from 1957 to 1974, and was only resolved in 2004 (Noguchi, 2014). The historical privileges enjoyed by the Shizuoka-based Sengen shrine, 'a religious (and thus private) organization' were vindicated in its property rights claim, as Japan's Supreme Court awarded it control over the summit of Fuji (Bernstein, 2013). However, the legal wrangles attracted considerable controversy, including opposition from organizations such as the National Parks Association and persistent criticism from others who were loath to accept the decision. To date, the ambiguous delineation of prefectural borders around the summit remains a poignant geographical reminder of the ownership issue. Nonetheless, the two prefectures have taken significant steps toward partnership in recent years, with waste counterstrategies the initial platform for collaboration. From 1967, both sides had begun to tackle the problem of rubbish left by tourists or dumped illegally on the mountainside and by 1980, both were organizing regular clean-up operations[7] (Sayama & Nishida, 2001) that symbolized the quest for WHS status.

Mount Fuji: from natural to cultural world heritage

The movement to build stakeholder collaboration accelerated after Japan ratified the UNESCO WHS convention in 1992, sparking speculation over Fuji's imminent listing (Oda, 2013). A 1991 Yamanashi Prefecture tourism management plan had announced the construction of a multi-storey car park at the mountain's foot, but the development was rejected after opposition from local conservation organizations. The campaign's success, together with the coincidental timing of the WHS ratification, encouraged a movement to have Fuji listed as a natural WHS. A petition had collected 2.4 million supporting signatures by 1994, when Fuji joined the first batch of destinations on Japan's tentative list (Noguchi, 2014). Subsequently, a joint council was launched by several central government agencies, with an interim report on conservation measures published in 1995 that paved the way for the set-up of the Fuji Charter and Country Code in 1998 (Sayama & Nishida, 2001). The secretariat was among the first in Japan to make formal provision for cross-cutting representation including bureaucrats from both prefectures, as well as the MOE and MAFF. This 'round-table' approach coincided with the collaborative management ideology that was assuming growing importance at national parks across the country (Tanaka, 2014), along with calls for more regular platforms for exchange of information or multi-stakeholder dialogue (Aikoh, 2014).

However, although Fuji remained on the tentative list as a natural WHS, feedback from the IUCN suggested that its likelihood of inscription was low due to: (i) the existence of other equivalent cone volcanoes, ensuring; (ii) unexceptional geological heritage as a stratovolcano; and (iii) its natural environment had been substantially altered by human activity, including; (iv) illegal dumping of waste (Noguchi, 2014, p. 74). This informal verdict by experts reflected the prevalence of 'ordinary' zoned land around the mountain's foot where planning permission laws had been powerless to prevent intense tourism infrastructure development such as resort hotels, toll roads, theme parks, golf courses and ski resorts (Hiwasaki, 2006). Ultimately, the proposal to apply for natural WHS listing was withdrawn in 2003 prior to the IUCN's official decision (Noguchi, 2014). Fuji's environmental management issues, symbolized by toilets and illegal waste, may not have been the principal stumbling block but they nonetheless contributed to the decision to abort the proposal for *natural* WHS status. Underpinning the debate was growing recognition of a vertically compartmentalized management model that had been singled out by a UNESCO report (1993) compiled during Yakushima's inscription as a natural WHS. It remained to be seen if listing could provide an external UNESCO 'carrot' to catalyze stakeholder collaboration at Fuji and across Japan's 'multi-purpose parks.'

Trail signs – pointing the way for collaboration?

Fuji's complex mix of land owners and managers was resulting in sectionalism that undermined the provision of visitor services and one tangible marker was the state of non-standard, broken, or redundant trail signs (Figure 2). After Fuji was reinstated on the tentative list in 2007, this time to apply for *cultural* WHS inscription, calls were made for a more holistic trail management plan beginning with new trail signs. In 2008, a petition was submitted to the respective prefectural governors requesting their standardization (Table 1). This prompted the set-up of a council, paving the way for information exchange between a cross-cutting, multi-tiered range of stakeholders including national, prefectural and municipal governments, together with private sector actors such as the mountain lodges. At the first meeting of the new council in 2009, the MOE was announced as the secretariat together with two prefectural governments (Appendix 1). The target area was defined as the upper section above the 5th station (2400 m ASL), together with major trails. The action plan for fiscal 2009 began in April with the set-up of a series of working groups. From July, the first new signs were set up on a trial basis, while a MOE survey plotted the dimensions and locations of all existing trail signs to propose future renovations (Figure 3).

The guidelines for holistic renovation of trail signs were announced provisionally at the second council meeting and finalized at the third meeting in 2010. They called for the standardization of place names in four languages, the creation of a standardized map, and the

Figure 2. Former trail signs reflect the complicated state of national park management (MOE, 2013).

Table 1. Timeline of trail sign council establishment.

24.09.2008	petition submitted to Prefectural Governors requesting the standardization of trail signs and set-up of a council
until March 2009	exchange of opinions between various stakeholders (MOE, both prefectures & municipal governments, mountain lodges etc.)
Sep 2008 - 09.03.2009	first meeting of newly-established council; overview of guidelines and timeline to complete determined within fiscal year 2009
April 2009	working groups set-up; summit working group by Yamanashi Prefecture; Shizuoka working group
July 2009	set-up of new signs in prime locations prior to start of climbing season
July-Sep 2009	MOE survey on types of trail signs plots all existing trail signs and proposes renovations
24.12.2009	2nd meeting of formal council releases draft documents: 'guidelines' and 'standardization plan'
23.03.2010	3rd meeting of formal council formalizes 'guidelines' and 'standardization plan'
30.06.2010	press announcement of 'guidelines' and 'standardization plan'
07.02.2011	4th meeting of formal council reports current state of standardization progress; results of climber questionnaire survey; follow-up to guidelines; future direction of council outlined → merged into broader council

removal of unnecessary or broken, signboards.[8] In this way, a system of color-coded signs with uniform design was introduced along the four main trails, creating a visual template for holistic management, and reducing visitor confusion such as at one junction on the return trail where climbers had often lost their way and descended on the wrong side of the mountain. The new signs had a multi-lingual format that reflected the increased presence of international climbers resulting partly from the heightened media presence

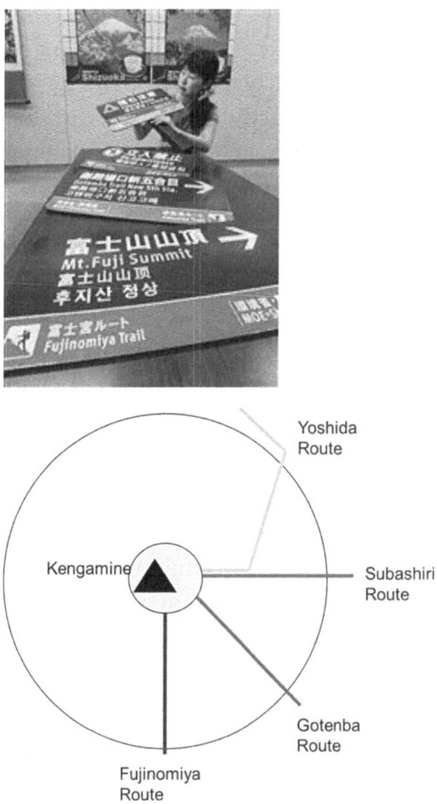

Figure 3. New color-coded trail signs introduced along the four routes from 2009.

which surrounded impending WHS listing. Although there is no explicit mention of the 'WHS factor' in the council's minutes, one of the academic experts present at the meeting noted the opportune timing and the participants' joint awareness of the likelihood that Fuji would shortly be listed. The new trail signs thus became a rallying point for stakeholder collaboration due to their simultaneous ability to offer a template for holistic management while providing utilitarian functions such as direction, distance, and warnings on behavior (Menke, 1996). The signs also provided a symbolic interface between park managers and visitors, representing a more pro-active approach. At the fourth meeting in 2011, future management directions were discussed and the council's operational mandate expanded to deal with a broader range of issues, including risk management, visitor services, environmental assessment, and interpretation of natural and cultural resources (Higashioka & Ieri, 2012). At the annual UNESCO convention in June 2013, Fuji was finally added to the WHS list under Criterions (iii): 'object of pilgrim worship' and (vi): 'source of artistic inspiration.'

From conflict to collaboration

This paper has investigated the hypothesis that Fuji's 2013WHS inscription helped usher in a new era of improved stakeholder collaboration, epitomized by a new system of trail signs. First, a literature review staked out the assertion that 'access is synonymous with conflict' in protected areas, depicted using Japan's national parks that rely on informal partnerships of public institutions such as the MOE and MAFF, together with local governments, private corporations and NGOs. To reduce conflict in this situation, the multi-landowner, multi- purpose system requires collaboration between all stakeholders to form 'partnership[s] to promote win–win solutions to achieving conservation goals' (Zinda, 2012). The task of coordinating those stakeholders falls predominantly to the MOE, yet their effectiveness is dependent on the cooperation of other actors, particularly the MAFF, which results in complex, multi-tiered decision-making arrangements. Our case study thus identified a range of administrative barriers that were undermining the capacity of Fuji's national park management.

First, park management plans include a zoning system empowering the MOE with specific regulatory powers to deny inappropriate development by granting or refusing planning permission; for example, by color coordinating building roofs in core zones. Yet zoning is strongly influenced by existing land ownership, with regulations retrofitted to reflect prior land use mosaics (Aikoh & Tomidokoro, 2010). It has thus proven difficult for the MOE to rein in development around Fuji's flanks, as was hinted at by the aborted natural WHS application, or even to outlaw vending machines from the summit crater which is classified as 'private' land owned by the shrine (Noguchi, 2014). In the face of diminished management effectiveness due to such sectionalism, trail signs became a rallying point for a new holistic template. The trail sign council's 'roundtable' format brought together the MOE and the two prefectures along with a cross-section of other stakeholders in a democratic consensus-building exercise. Unlike the conventional 'command-and-control' style (Barrett & Therivel, 1991), the more inclusive management model could help counter the 'vertical compartmentalization' criticized by UNESCO (1993) at Yakushima.[9] Fuji's trail signs signify an innovative approach by creating an effective partnership using 'consensus-based and collaborative decisions' (Hamin,

2001). Unlocking the potential of multi-use parks requires just this kind of cross-cutting interaction, with the capacity for targeted management intervention. The trail signs represent a standardized, multi-lingual response, although they still face practical challenges such as the descending trail junction along the Shizuoka-Yamanashi border where the new signs were unable to prevent 968 reported cases of 'lost' climbers in 2013. Trail sign content is another issue, with the lack of cultural interpretation belying the WHS categories that formed the crux of the inscription. Within the trail sign council itself, there is still need for a more transparent decision-making process that includes a role for neglected actors such as the members of civil society and the tourism sector. The basic requirements of the collaborative approach such as openness, integration and co-operation are otherwise undermined by a park management style which offers few opportunities for exchange of information and makes no legal provision for bottom-up participation in the decision-making process (Aikoh, 2014; Hiwasaki, 2005). The council has rendered a cross-section of stakeholders a formal role, but there remains insufficient discussion on how best to then involve them in collaborative decision-making (Webler, Tuler, & Tanguay, 2004). Like Konisky and Beierle's (2001) verdict on public comments and hearings, the council may be 'reactive in nature, involve insufficient deliberation, and engage only a small number of participants' without careful nurturing.

The Fuji example also highlights the constant struggle between the two parallel acts of responding to international diplomacy while offsetting domestic political agenda. Although UNESCO has no formal jurisdiction over WHSs, the required Management Plan offers an opportunity to shape domestic policy-making by disseminating core values such as participation as a platform for increased efficiency (Webler et al., 2004). Fuji's trail signs could be one indirect example but the question is how to maintain momentum in the aftermath of listing. Herein lies the 'double-edged sword' of WHS listing as an external 'carrot' to build consensus. In fact, 'UNESCO has many carrots but hardly any sticks. It is strongest during the process of evaluation, when it can decide whether or not to grant World Heritage status to a site and on what conditions' (Avieli, 2015). Building site-level capability through a combination of rigorous requirements followed up with regular evaluation would thus be in UNESCO's own interests, in light of the 38 WHSs on the endangered list and two sites that have been struck off due to lack of compliance.[10] Fuji offers a final warning of potential post-listing pitfalls, as the two prefectural governments that were seen to have synchronized the date of their mountain opening days for the start of the 2013 season, then reverted to different dates in 2014 after WHS inscription (Noguchi, 2014).

Limitations and future research

The methodological approach applied in this study is not without its limitations. Since a comprehensive inventory analysis of all the actors involved was deemed unfeasible, two central and two local government actors were selected for in-depth analysis. Their selection was justified in that they were all found to play a pivotal role in the overall administration of Japan's national parks (Knight, 2010). However, in the case of Fuji it is the local stakeholders, particularly the prefectural governments, whose decision-making came to the fore (Noguchi, 2014). A comparative study with other national parks, particularly one such as Shiretoko with a high proportion of national forest, could be more insightful in mapping the contours of the unresolved overlap between the MOE and MAFF administrations. However, Fuji's highly complex list of the stakeholders (Appendix 1) does provide a window into the practical difficulties that surround the promotion of a collaborative management approach. The challenge now facing 'multi-purpose' protected area managers is how to provide consistent objectives and targeted services based around a holistic vision. In Japan – as elsewhere – this is exacerbated by the lack of a single land owner or manager, coupled with irregular opportunities for participation (Aikoh, 2014). By pinpointing the complicated state of Japan's multi-purpose national parks, this paper adds to the growing body of research on stakeholder collaboration in protected areas and gives fresh insight into the WHS program's potential to become an external conflict-resolution mechanism which helps catalyze the creation of site-level partnerships.

Notes

1. Originally passed in 1957 and amended in 1970 following the establishment of the Environment Agency (Imura, 2005).
2. The MOE's land ownership amounts to a mere 0.2% of all national parks in Japan (Katō, 2008, p. 23).
3. Including the three previously listed *natural* WHSs of Shiretoko, Shirakami and Yakushima, but not Ogasawara which was added to the list in 2013.
4. For example, the number of people engaged in forestry decreased from 450,000 (1960) to 50,000 (2005), while the ratio of those aged over 65 increased from 5% to 27% (Matsumoto, 2013).
5. The MLIT emerged from a 2001 amalgamation of the Construction and Transport Ministries that was intended to further decentralization but 'actually had the opposite effect' (Feldhoff, 2002).
6. This amounted to some JPY10 trillion, equivalent to US$100 billion in FY2000 (Feldhoff, 2002).
7. The Fuji-san Oyobi Shūhen Bika Seisō Katsudō in Yamanashi and the Fuji-san wo Itsumade mo Utsukushiku Suru Kai in Shizuoka Prefectures respectively.
8. http://www.pref.shizuoka.jp/a_content/trend/2009_07_03.html.
9. An island located between Kyushu and Okinawa listed as a *natural* WHS in 1993.
10. Oman's Arabian Oryx Sanctuary in 2007 and Dresden's Elbe River Valley in 2009.

Acknowledgements

Thanks to AK, KY and the PREC Institute Inc. Also to many other professional contacts and peers who contributed to this paper, and to JSPS Grant No. 26760023 and 15H03274.

Disclosure statement

No potential conflict of interest was reported by the authors.

References

Aikoh, T. (2014). Issues in planning and management in the Japanese national park system: Perspectives from Daisetsuzan National Park. *The Japanese Forest Economic Society*, *60*(1), 14–21. In Japanese.

Aikoh, T., & Tomidokoro, Y. (2010). Relationship between the national park plan and the national forest plan in Shikotsu-Toya National Park. *Journal of the Japanese Institute of Landscape Architecture, 73*(5), 505–508. In Japanese.

Avieli, N. (2015). The rise and fall (?) of Hoi An, a UNESCO world heritage site in Vietnam. *SOJOURN: Journal of Social Issues in Southeast Asia, 30*(1), 35–71.

Axelrod, R. (1997). The dissemination of culture - A model with local convergence and global polarization. *Journal of Conflict Resolution, 41*(2), 203–226.

Barrett, B. F. D., & Therivel, R. (1991). *Environmental policy and impact assessment in Japan.* London: Routledge.

Beeton, S. (2002). *Eko Tsu-Rizumu Kyouhon.* Tokyo: Heibonsha Ltd.

Becton, S. (2006). *Community development through tourism.* Melbourne: Landlinks Press.

Bernstein, A. (2013). *Guns and grass: The militarization of Fuji's common land.* Paper delivered at annual meeting of The International Association for the Study of the Commons, June 2013.

Blahna, D. J., & Yonts-Shepard, S. (1989). Public involvement in resource planning: Toward bridging the gap between policy and implementation. *Society & Natural Resources, 2*, 209–227.

Buckley, R. (2004). The effects of world heritage listing on tourism to Australian national parks. *Journal of Sustainable Tourism, 12*(1), 70–84.

Cawley, M., & Gillmor, D. A. (2008). Integrated rural tourism: Concepts and practice. *Annals of Tourism Research, 35*(2), 316–337.

Chakraborty, A., & Jones, T. (2018). Mount Fuji: The volcano, the heritage and the mountain. In A. Chakraborty, K. Mokudai, M. Cooper, M. Watanabe, & S. Chakraborty (Eds.), *Natural heritage of Japan: Geological, geomorphological, and ecological aspects* (pp. 167–176). Switzerland: Springer.

Curry, N. R. (1994). *Countryside recreation, access and land use planning.* London: E and FN Spon.

Eagles, P. (2002). Trends in park tourism: Economics, finance and management. *Journal of Sustainable Tourism, 10*, 132–153.

Evans, N. (2002). Machi-zukuri as a new paradigm in Japanese urban planning: Reality or myth? *Japan Forum, 14*, 443–464.

Feldhoff, T. (2002). Japan's regional airports: Conflicting national, regional and local interests. *Journal of Transport Geography, 10*(3), 165–175.

Fennell, D., & Weaver, D. (2005). The ecotourium concept and tourism-conservation symbiosis. *Journal of Sustainable Tourism, 13*(4), 373–390.

Foljanty-Jost, G. (2005). NGOs in environmental networks in Germany and Japan: The question of power and influence. *Social Science Japan Journal, 8*(1), 103–117.

Forestry Agency. (2006). *Forestry white paper.* Tokyo: MAFF. In Japanese.

Hamin, E. (2001). The US national park service's partnership parks: Collaborative responses to middle landscapes. *Land Use Policy, 18*, 123–135.

Hammer, T., Mose, I., Siegrist, D., & Weixlbaumer, N. (2007). Protected areas and regional development in Europe: Towards a new model for the 21st century. In I. Mose (Ed.), *Protected areas and regional development in Europe: Towards a new model for the 21st century* (pp. 233–246). Farnham: Ashgate.

Haukeland, J. V. (2011). Tourism stakeholders' perceptions of national park management in Norway. *Journal of Sustainable Tourism, 19*(2), 133–153.

Heritier, S. (2010). Public participation and environmental management in mountain national parks. *Journal of Alpine Research, 98*, 170–188.

Higashioka, R., & Ieri, K. (2012). Fuji no riyou tekiseika ni muketa kyoudou no torikumi. *Collaborative Efforts Toward Appropriate Visitor use of Mt Fuji). National Parks, 705*, 7–10. In Japanese.

Hiwasaki, L. (2005). Toward sustainable management of national parks in Japan: Securing local community and stakeholder participation. *Environmental Management, 35*(6), 753–764.

Hiwasaki, L. (2006). Community-based tourism: A pathway to sustainability for Japan's protected areas. *Society & Natural Resources: An International Journal, 19*(8), 675–692.

Huang, C.-H., Tsaur, J.-R., & Yang, C.-H. (2012). Does world heritage list really induce more tourists? Evidence from Macau. *Tourism Management, 33*, 1450–1457.

Imura, H. (2005). Japan's environmental policy: Institutions and the interplay of actors. In H. Imura, & M. A. Schreuers (Eds.), *Environmental policy in Japan* (pp. 49–85). Cheltenham: Edward Elgar.

Ishikawa, T. (2001). *Nihon no shizen-hogo*. Tokyo: Heibonsha. In Japanese.

Ito, T. (1996). Influence of forestry on the formation of national park policy in Japan. *Japanese Journal of Forest Planning, 2*, 85–95.

Jamal, T., & Getz, D. (1995). Collaboration theory and community tourism planning. *Annals of Tourism Research, 22*, 186–204.

Jimura, T. (2011). The impact of world heritage site designation on local communities – A case study of Ogimachi, Shirakawa-mura, Japan. *Tourism Management, 32*(2), 288–296. doi:10.1016/j.tourman.2010.02.005

Jones, T. E. (2015). Hokkaido's protected areas and regional revitalization: The case study of Toya-Usu Geopark in Shikotsu-Toya National Park. In S. Assmann (Ed.), *Sustainability in contemporary rural Japan: Challenges and opportunities* (pp. 151–166). London: Routledge.

Jones, T. E., Yamamoto, K., & Aramaki, S. (2012). *Building a proactive conservation strategy for Mount Fuji's foreign climbers*. 6th international conference on monitoring and management of visitors in recreational and protected areas, Stockholm, Sweden, August 21–24.

Jones, T. E., Yang, Y., & Yamamoto, K. (2018). Comparing automated and manual visitor monitoring methods: Integrating parallel datasets on Mount Fuji's north face. *Journal of Park and Recreation Administration, 36*(1), 22–38.

Katō, M. (2008). *National park system of Japan, national park & protected area management series (Vol. III)*. Tokyo: Kokon Shoin. In Japanese.

Knight, C. (2010). Natural environments, wildlife, and conservation in Japan. *The Asia-Pacific Journal, 4*, 2–10.

Konisky, D. M., & Beierle, T. C. (2001). Innovations in public participation and environmental decision making: Examples from the Great Lakes region. *Society & Natural Resources, 14*, 815–826.

Kureha, M., Kikuchi, T., Sano, M., & Yamamoto, M. (2015). Overview of special issue: Human geographical studies of Mount Fuji region focusing on tourism. *Journal of Geography, 124*(6), 877–885. In Japanese.

Kuvan, Y., & Akan, P. (2012). Conflict and agreement in stakeholder attitudes: Residents' and hotel managers' views of tourism impacts and forest-related tourism development. *Journal of Sustainable Tourism, 20*(4), 571–584.

MacMillan, D. C., & Phillip, S. (2008). Consumptive and non-consumptive values of wild mammals in Britain. *Mammal Review, 38*, 189–204.

Mahanty, S., & Russell, D. (2011). High stakes: Lessons from stakeholder groups in the biodiversity conservation network. *Society & Natural Resources, 15*(2), 179–188. doi:10.1080/089419202753403346

Mason, R. J. (1999). Whither Japan's environmental movement? An assessment of problems and prospects at the national level. *Pacific Affairs, 72*(2), 187–207. doi:10.2307/2672119

Matsumoto, Y. (2013). *Masters of forests: Their wisdom and arts*. Paper presented at APC on November 15 2013. Rural Development and Greening Office, Forestry Agency, Japan.

Menke, B. (1996). *Handbook for trail design, construction & maintenance*. US National Parks Service. Website. Retrieved from http://www.nps.gov/noco/parkmgmt/ncttrailconstructionmanual1.htm

Ministry of the Environment (MOE), Government of Japan. (2013). *Good practices in the management of national parks and other protected areas in Japan* (pp. 68–69). Tokyo: MOE.

Mitchell-Banks, P. (1997). Community, culture and conflict. In B. Solberg & S. Miina (Eds.), *Conflict management and public participation in land management* (1997th ed., Vol. *14*, pp. 149–164). Joensuu: The European Forest Institute.

Mitsuda, H., & Geisler, C. (1992). Imperilled parks and imperilled people: Lessons from Japan's Shiretoko national park. *Environmental History Review, 16*(2), 23–39.

Morgan-Davies, C., & Waterhouse, A. (2010). Future of the hills of Scotland: Stakeholders' preferences for policy priorities. *Land Use Policy, 27*, 387–398.

Murakushi, N. (2005). *The history of the establishment of Japan's National Parks* (p. 155). Tokyo: Housei University Press. In Japanese.
Natori, Y. (1997). Shiretoko logging controversy: A case study in Japanese environmentalism and nature conservation system. *Society & Natural Resources, 10*, 551–565.
Noguchi, K. (2014). *Sekai isan ni sarete Fuji ha naite iru.* Tokyo: PHP Research. In Japanese.
Oda, Z. (2013). *Naze Fuji ha sekai isan ni natta no ka.* Tokyo: PHP Research. In Japanese.
Oyadomari, M. (1989). The rise and fall of the nature conservation movement in Japan in relation to some cultural values. *Environmental Management, 13*(1), 23–33.
Palau-Saumell, R., Forgas-Coll, S., Sánchez-García, J., & Prats-Planagumà, L. (2013). Tourist behavior intentions and the moderator effect of knowledge of UNESCO world heritage sites. *Journal of Travel Research, 52*(3), 364–376.
Pascual, U., & Perrings, C. (2007). Developing incentives and economic mechanisms for in situ biodiversity conservation in agricultural landscapes. *Agriculture, Ecosystems & Environment, 121*(3), 256–268.
Polidor, A. (2007). *Sacred site reports.* Retrieved from http://www.sacredland.org/mount-fuji
Poria, Y., Reichel, A., & Cohen, R. (2011). World heritage site-is it an effective brand name?: A case study of a religious heritage site. *Journal of Travel Research, 50*(5), 482–495.
Ryan, J., & Silvanto, S. (2009). The world heritage list: The making and management of a brand. *Place Branding and Public Diplomacy, 5*(4), 290–300.
Sautter, E. T., & Leisen, B. (1999). Managing stakeholders: A tourism planning model. *Annals of Tourism Research, 26*(2), 312–328.
Sayama, H., & Nishida, M. (2001). Transition of the beautification and cleaning activities in Mt. Fuji area after World War II. *Landscape Research Japan, 64*(5), 485–488. In Japanese.
Schreurs, M. (2005). Japan and global environmental governance. In G. Hook (Ed.), *Contested governance in Japan: Sites and issues* (pp. 157–175). London: Routledge.
Seaton, P. (2010). Depopulation and financial collapse in Yubari: Market forces, administrative folly, or a warning to others? *Social Science Japan Journal, 13*(2), 227–240.
Sharpley, R. (2009). The English Lake district - national park or playground? In W. Frost & C. M. Hall (Eds.), *Tourism and national parks: International perspectives on development, histories, and change* (pp. 155–166). Oxon: Routledge.
Stewart, J. M. P., & Sinclair, A. J. (2007). Meaningful public participation in environmental assessment: Perspectives from Canadian participants, proponents, and government. *Journal of Environmental Assessment, 9*(2), 161–183.
Tanaka, T. (2014). Governance of Japan's national parks in 21st century: A perspective from public administration. *Landscape Research Japan, 78*(3), 226–229. In Japanese.
Totman, C. (2007). *Japan's imperial forest, Goryōrin, 1889–1946: With a supporting study of the kan/min division of woodland in early Meiji Japan, 1871–76.* Folkestone: Global Oriental.
Tsuchiya, T. (1981). Tourism development through investment in transport in the prewar period; a case study of Fujikyūkō. *Japanese Journal of Forest Economics, 100*, 30–34. In Japanese.
United Nations Educational, Scientific and Cultural Organization (UNESCO). (1993). *Periodic reporting for well-planned heritage preservation.* Retrieved from whc.unesco.org
Watanabe, T., Sasaki, S., Shinohe, H., & Shimomura, A. (2012). A study on the transition of resource values of the national parks in Japan and the management style. *Landscape Research Japan, 75*(5), 483–488. In Japanese.
Webler, T., Tuler, S., & Tanguay, J. (2004). Competing perspectives on public participation in national park service planning: The Boston Harbor Islands national park area. *Journal of Park and Recreation Administration, 22*, 91–113.
Wilson, E., Nielsen, N., & Buultjens, J. (2009). From lessees to partners: Exploring tourism public–private partnerships within the New South Wales national parks and wildlife service. *Journal of Sustainable Tourism, 17*(2), 269–285. doi:10.1080/09669580802495774
Yin, R. K. (2009). *Case study research: Design and methods (applied social research methods).* London: SAGE.
Zinda, J. A. (2012). Hazards of collaboration: Local state co-optation of a new protected-area model in southwest China. *Society & Natural Resources, 25*(4), 384–399.

Appendix 1. Table of stakeholders involved in trail sign council.

Ministry of Education, Culture, Sports, Science & Technology, Agency for Cultural Affairs
Ministry of Agriculture, Forestry and Fisheries, Kanto Regional Forest Office
Ministry of Land, Infrastructure, Transport and Tourism, Chubu Regional Development Bureau
Ministry of Land, Infrastructure, Transport and Tourism, Chubu Regional Development Bureau
Ministry of the Environment, Kanto Regional Environment Office
Ministry of Defense, JGSDF Fuji School
Ministry of Defense, JGSDF Camp Kita-Fuji
Yamanashi Prefecture, Tourism Depertment
Yamanashi Prefecture, Prefectural Land Development Department
Yamanashi Prefecture, Forestry and Environment Department
Yamanashi Prefecture, Planning and Resident Life Department
Yamanashi Prefecture, General Affairs Department
Yamanashi Prefecture, Board of Education
Yamanashi Prefecture, Police Department
Fujigoko Fire Department
Fuji-Yoshida City
Fuji-Kawaguchiko Town
Narusawa Village
Onshirin Regional Public Organization
Narusawa/Fuji-Kawaguchiko Onshirin
Yamanashi Prefecture Highway Corporation
Shizuoka Prefecture Community and Environmental Affairs Department
Shizuoka Prefecture Emergency Management Department
Shizuoka Prefecture Culture & Tourism Department
Shizuoka Prefecture Transportation Infrastructure Department
Shizuoka Prefecture Board of Education
Shizuoka Prefecture Police Department
Fujinomiya City
Gotemba City
Oyama City
Fuji City
Susono City
Fujisan Hongū Sengen Taisha
Fuji Five Lakes Tourism Association
Mt Fuji Yoshidaguchi Innkeepers Association

Repertory grids and the measurement of levels of community support for rural ecotourism development

Stephen Schweinsberg, Simon Darcy and Stephen L. Wearing

ABSTRACT

Many of the world's iconic ecotourism attractions reside in heterogeneous rural localities (or places). Over time the nature of these settings evolve, which can lead to tourism and other industry sectors (e.g. forestry, mining, agriculture etc.) at different times coming to hold positions of legitimacy in the eyes of local people. Local people will form opinions on the merits of ecotourism development on the basis of an evolving interplay of economic, environmental and social forces. These forces are framed on the basis of a locality's history, as well as on the basis of current conditions. In the present paper the authors will explore the merits of employing Personal Construct Theory based repertory grids as a methodological tool to understand the subjective realities of local people.

Drawing on the results an expanding body of scholarship that has used repertory grid methodologies in destination image studies; the authors will offer comment on the merits whereby repertory grids can be taken out of an academic led research setting and used within communities themselves to shed light on the individual perceptions that variously cause support or opposition to ecotourism development.

Introduction

In 1974 the French novelist, filmmaker and documentalist Georges Perec wrote prophetically that the 'country is a foreign land. It shouldn't be, yet it is' (Perec, 1997, p. 68). Written as part of a series or personal reflections on human society's relationship with space, Perec's reflections on the countryside contain a number of observations that will strike a chord with ecotourism planners. Firstly, as Perec (1997) notes the countryside is for many urbanites a decorative space; a space they observe passively as a means of escape from the drudgery of their daily existence. Early conceptualizations of ecotourism, which grew out the environmental movement (e.g. Ceballos-Lascurain, 1987), were concerned primarily with articulating those opportunities that existed for ecotourists to interact with a passive natural and cultural setting. Whilst later definitions of ecotourism now encompass a more overt ethical and sustainable development imperative that includes greater focus on socio-cultural and other variables (Donohoe & Needham, 2006;

Weaver, 2005), Cater (2006) has identified an increasing trend towards the institutionalization of ecotourism and the grounding of ecotourism discourses in what are described as 'Western cultural, economic and political processes' (Cater, 2006, p. 23). The potential for such processes to figuratively 'freeze out' host communities in particular destination regions and create white spaces (see O'Doherty, De Cock, Rehn, & Ashcraft, 2013) or foreign lands is the raison de'étre for the present paper.

While community impacts have long been recognized as an important aspect of ecotourism scholarship (Duffy, Kline, Swanson, Best, & McKinnon, 2017; Stronza & Gordillo, 2008); Cobbinah (2015) has observed that it was only in the last decade or so that ecotourism's remit is generally seen to have extended beyond environmental conservation to encompass the socio-cultural concerns of a diverse stakeholder population. While community based ecotourism operations and associated scholarship has continued to grow (e.g. Jones, 2005; Stone, 2015) Liu et al. (2014) have observed that cognitive social capital functions as one of the primary determinants of a community member's level of pro environmental behaviour. Cognitive social capital refers to 'values, attitudes, norms and beliefs' that are prevalent in host communities (Liu et al., 2014, p. 191). In this paper we will seek to explore the merits in using repertory grid methodologies to shed light on these subjective local realities. Repertory grids have already been employed in a range of destination image studies (Chang & Mak, 2018; Yeh & Cheng, 2015). Pike and Kotsi (2016) have argued in favour of the method, noting that by imbuing tourism stakeholders (in our present case residents) with the power to construct individual interpretations of the world that align to their own individual experience we are better able to understand the breadth of perspectives that may exist in a community of interest. In the next section we will discuss aspects of the use of Personal Construct Theory based repertory grid methods in tourism scholarship. This will then be followed in the final section with commentary on the merits of using repertory grids to understand subjective community realities in contested or wicked ecotourism settings.

A psychology lens, repertory grids and tourism

Ever since the pioneering work of academics including Phillip Pearce and Peter Stringer in the 1980s and 1990s (see Pearce, 2013; Pearce & Stringer, 1991; Stringer & Pearce, 1984) psychology has come to be seen along with geography and sociology as one of the underlying disciplinary foundations for the study of tourism (Tribe, 1997). Pearce and Packer (2013) have in that time observed a shift in how psychology – tourism links are fostered, noting a departure from pioneering work on individual psychoanalysis and behaviourism and a renewed focus on topics like cultural context and its effect on tourism development. As Pearce and Packer (2013, p. 405) have observed 'psychology has acknowledged its own cultural relativism … Whether they are culturally tied or pan-cultural, tourism studies [similarly] need to acknowledge the important influence of cultural context and in doing so might enhance psychological inquiry'. Pearce and Pearce (2017) have recently sought to articulate how the evolution in the nature of tourism inquiry to encompass concern for subjective cultural context (e.g. the critical turn, sustainability) have necessitated the development of new research methodologies. Research subjects, they note will not always be cognizant or equipped to partake in traditional surveys and other data gathering processes. Pursuing techniques, which they note 'rely less on verbal fluency with the

Western concept of a survey can help access the views of these study audiences' (Pearce & Pearce, 2017, p. 19).

Personal Construct Theory (hereafter PCT) was developed by the American Psychologist George Kelly in the mid twentieth century with the key tenants of the theory being popularized in the work *A Theory of Personality: The Psychology of Personal Constructs* (Kelly, 1963). Developed initially for application in the area of clinical psychology, the central postulate of PCT says that a 'person's processes are psychologically channelized by the way in which he (sic) anticipates events' (Kelly, 1963, p. 46). By proposing that human beings are able to interpret their environment on the basis of a series of self-generated hypotheses on how the world works, PCT and repertory grids have proven to be a valuable theoretical and methodological lens for interpreting a range of environmental and tourism management issues. Tourism PCT applications have included: destination image analysis studies (Botterill & Crompton, 1987; Coshall, 1991, 2000; Embacher & Buttle, 1989; Hankinson, 2004a, 2004b; Jenkins, 1999; Naoia, Aireyb, Iijimaa, & Niininenc, 2006; Pike, 2012; Pike & Ryan, 2004; Walmsley & Jenkins, 1993); tourist perceptions of nature (Waitt, Lane, & Head, 2003); the formulation of the tourist experience (Botterill & Crompton, 1996; Scherl, 1988); tourist behaviours (Mak, Lumbers, Eves, & Chang, 2013); destination planning (Yeh & Cheng, 2015); host perspectives of tourism development (Mansfield & Ginosar, 1994; Schweinsberg, Wearing, & Darcy, 2012); and host perspectives on changes to the nature of forest and rural environments (Coakes, Fenton, & Gabriel, 1999; Wearing, Schweinsberg, Lai, & Lyons, 2014).

To begin to see why PCT has gained such an important place in the study of tourism we must look to both the originator of the approach and the method he developed (the repertory test and repertory grid analysis). In 1979 Kelly wrote a paper, which he entitled 'Ontological Acceleration' in which he sought to understand more about the anticipatory nature of human behaviour (McWilliams, 2008). By proposing that human behaviour is not simply the result of past causes, which McWilliams (2008, p. 60) notes would entail some 'regularity of relations between antecedent events and behavior'; human behaviour is instead seen as active, independent of the will of policy makers and planners. The importance of developing an epistemological position, which accounts for the active creation of tourist places, is a recognized concern of the literature. Young (1999, p. 373) notes that 'ordinary places become tourist places when they are attributed particular meanings and values which appeal to tourists'. Similarly, rural areas are subject to a range of competing stakeholder viewpoints. Halfacree (2007) defines the contestation between viewpoints as a rural region's Trial by Space. The idea of a Trial by Space was modelled by Halfacree off earlier work the French Philosopher Henri Lefebvre (see Lefebvre, 1991 [1974]) and proposes that 'value acquires or loses its distinctiveness through confrontation with other values and ideas that it encounters'. Elden (2004, p. 185) talks of trial by space as being associated with the way that space is 'perpetually recast' by people. The notion of constructive alternativism, which underpins PCT, proposes that 'all of our present interpretations of the universe are subject to revision and replacement' (Kelly, 1963, p. 15). To understand how these revisions are formulated requires the development of emic methodologies, methodologies that articulate the way that specific cultures express the nature of their activities (see Niblo & Jackson, 2004). It is with this idea that we now turn our attention to repertory grid methodologies.[1]

Repertory tests and repertory grids analysis represent the outcome of a structured interview (and sometimes survey) process whereby an individual's personal constructions can be elicited and later mapped and interpreted by a researcher (see Fransella, Bell, & Bannister, 2004). Repertory tests refer to the process of eliciting 'an individual's range of constructs used to guide their behaviour in a certain context' (Pike, 2018, p. 223). Repertory grids, in contrast, 'can be used to examine the construct system of one individual' (Pike, 2018, p. 223). Underpinning the methodologies are two concepts, elements (or topics of inquiry) and constructs (an individual's interpretation of the meaning they attach to the relationship between certain elements). Elements are typically presented to respondents in a triad fashion, which in the context of urban tourism was described by Pearce and Pearce (2017, p. 24) in the following terms –

> Consider the cities of Hong Kong, London and Rome ... in what ways are London and Hong Kong alike but different from Rome? Next, in what ways are Rome and Hong Kong alike but different from London, and finally in what ways are London and Rome alike but different from Hong Kong?

When element triads are presented and constructs elicited, Home, Bauer, and Hunziker (2007, pp. 49–50) describe their relationship in the following terms:

> One of the central assumptions is that constructs are dichotomous and built from contrasts rather than absolutes. An element ... will receive meaning by it being seen as both that which it is and contrasted with that which it is not. However, in expressing a meaning, an element is evaluated within a contrast rather than a negative. For example the meaning intended by the descriptor 'attractive', can best be understood when opposed to its contrast which may be say 'disinteresting', 'ugly' or 'repulsive'. Accordingly, learning constructs with both poles described, allows the researcher to gain a fuller understanding of how a person construes an element than if only a descriptor was found.

Pearce and Pearce (2017) have argued that the value in constructs formed through triad based repertory grid elicitation can be seen at multiple levels. On the one hand constructs represent novel information on the perceived nature of the respondent's world, information that can be used to shed light on the forces that are likely to shape attitude and perception. At a higher level, constructs can then be used as the foundation of later interview and survey work (Pearce & Pearce, 2017). In doing so the words of real people become the basis of psychological questioning. With no need for coding or artificial adjustment on behalf of the researcher, the completed repertory grid with its lists of constructs plotted in relation to the comparison of different combinations of elements becomes a mechanism for shining a light onto previously inaccessible sources of knowledge and thus power in contested place based settings.

Schweinsberg, Wearing, and Wearing (2015) have observed that the transformation of an area through tourism development will seldom lead to lead to homogenization of culture. Instead they note, communities will create a sense of belonging and place 'through cultural and social interactions, personal experiences and individual actions and beliefs' (Schweinsberg et al., 2015, p. 105). In PCT, individuals are the prime source of their own constructions, albeit influenced by a range of localized relationship and social factors that may have the effect of creating 'hostility' to individuals who maintain broken constructions even in the face of invalidating evidence (see Kelly, 1969). As a technique that bridges traditional qualitative and quantitative divides (Fransella et al.,

2004), repertory grids are well positioned to provide researchers with a structured mechanism for engaging with the inherent messiness of engaging with marginalized voices in tourism destination settings. To date, there has been limited application of such methods to the study of host populations in contested or wicked tourism settings (notable exceptions include Lawton, 2005; Mansfield & Ginosar, 1994; Schweinsberg et al., 2012).

Repertory grids and the wicked complexity of an ecotourism setting

Ecotourism frequently exists in a heterogeneous setting where the industry is subject the effects of a diverse rage of stakeholder influences (e.g. Schweinsberg, Darcy, & Cheng, 2017; Wearing, Schweinsberg, & Tower, 2016). Where an issue polarizes a social group it becomes a wicked problem in that 'various stakeholders will disagree over [how the issue should be framed and its] … resolution' (Coombs & Holladay, 2018, p. 79). Initially defined by Churchman (1967, p. 141) as a class of social problems 'which are ill formulated, where the information is confusing, where there are many clients and decision makers with conflicting values, and where the ramifications in the whole system are thoroughly confusing'; the term then came to be popularized in the work of Rittel and Webber (1973) into the intractability of stakeholder perceptions on urban land use policy formulation. Since then the lens of wicked problems has been applied to a wide range of management contexts, including: strategic management (Camillus, 2008), fisheries management (Jentoft & Chuenpagdee, 2009; Khan & Neis, 2010), rural policy and indigenous governance (Christie, Rowe, & Pickernell, 2009; Head, 2008; Johns, 2008) and climate change (Lazarus, 2008; Levin, Cashore, Bernstein, & Auld, 2012).

To date there has been limited specific canvassing of wicked problem philosophies in a tourism management context (exceptions include Brooks & Champ, 2006; McCool, Patterson, Gartner, & Lime, 2000; Odeh & Diaz, 2013; Shultis & Way, 2006). While external influences on tourism like climate change are described as super wicked problems (Hall, Gössling, & Scott, 2015; Levin et al., 2012) actual applications of wicked problem mindsets to the study of sustainability and alternative tourism forms are rare. Scherrer and Doohan's (2014) investigation of tourism access to sensitive Aboriginal sites on Australia's northern Kimberley coast identifies both the cultural and historical framing of wicked problems. Wickedness is not a statement of sinister intent but rather a statement of social complexity and the intractability of different value positions. Scherrer and Doohan (2014) observe that the ability of tourism interests to take ownership of a wicked problem will be dependent initially upon a realization that different social perspectives exist, essentially a realization that the problem is wicked. When such complexities are recognized, Thompson and Whyte (2012) note we can then recognize the presence of

> multiple value perspectives and undertake inquiries and discussions with the aim of learning how to reconstruct the various wicked dimensions, from scientific uncertainty to disagreement over problem formulation, in ways that reveal possibilities for ameliorative or adaptive responses. (p. 489)

Repertory grids offer the opportunity to frame wicked problem dimensions that have relevance to them. Writing in a destination image context, Pike (2018) has recently argued that PCT and its associated methodologies have provided a qualitative method for identifying 'destination attributes salient to consumers' (p. 227). Similarly, Schweinsberg et al. (2012) undertook what was, at the time, their first foray into the employment of repertory grid methodologies to measure the community voice in contested rural tourism locales (see Schweinsberg et al., 2012). From this paper it was concluded that repertory grids do offer planners the opportunity to invest in local people themselves with the power to frame debates, as opposed to externally focused social impact assessors. What haven't been explored are the practicalities of taking repertory grid formation out of the hands of academics and planners and investing responsibility for its formulation solely in the hands of local people. In the following paragraphs we will offer some observations on this question, drawing both from our own experience and the literature.

When Kelly (1963) proposed his original PCT treatise he included the now famous metaphor of 'man the scientist' (*sic*). There was inherent in this metaphor an appreciation by Kelly of the communal and empowering nature of the theory he was proposing. No longer were experts to be identified on the basis of a particular class. Instead Kelly (1963) was proposing that all human beings should be seen as having the power to construe the impact of changes in their situation. As (Kawaf & Tagg, 2017, p. 225) observe, by 'viewing man as a scientist and [thus] as an expert in his/ her experiences ... [the same individual becomes] the best source to guide us into understanding his/ her experiences n holistic manner'. While the repertory grid therefore has a lot of merit in allowing individuals to test and revise their perceptions on account of their own experience; the usability of the procedure outside of the sphere of academia is open to question. With this in mind we will conclude by briefly considering some of the challenges that would be involved in the development of an adaptable repertory grid tool that can be employed by local people with limited external engagement with outside experts. Our focus here will be at the most basic stages of the process – approach validity, element selection, methodological explanation and construct formation (see also Bell, 1997; Fransella et al., 2004; Jankowicz, 2004; Pike, 2007; Potter & Coshall, 1984; Stewart, 2004).

Garrod once observed that whilst 'ecotourism is increasingly being advanced as a strategy to help address economic and social problems in local communities ... the full and effective participation of local communities in planning and management ... is, however, rarely a feature of ecotourism projects' (Garrod, 2003, p. 33). In the context of protected areas in Malawi, Bello, Lovelock, and Carr (2017, p. 131) have identified a range of factors constraining community participation including: 'apathy; inadequate financial resources; inadequate information; low education levels; unfair distribution of benefits; lack of trained human resources; a centralized public administration; lack of coordination; and human–wildlife conflicts'. Bello et al. (2017) called for the development of approaches or strategies that can be used to facilitate community participation. The challenge with achieving such a goal is, however, that methods must be aligned to the communities in which they are to be employed (see Reggers, Grabowski, Wearing, Chatterton, & Schweinsberg, 2016). At its most basic level, alignment of repertory grid methodologies to community interests must consider – what is the community members' knowledge and perception of the repertory grid?

While the repertory grid technique has been around in different forms for half a century, Pike (2018) has acknowledged that globally there is only a relatively small active research community. Anecdotal evidence suggests that research subjects often find repertory grids interesting and a good method for 'helping them to understand aspects of their decision making that they had previously not recognised' (Pike, 2018, p. 228). At the same time, however, other studies have drawn attention to the 'foreign' nature of the element/ construct elicitation process (Canning & Holmes, 2007). From our own experience, local residents on account of what are often cynical perspectives on the merits of academic research want to immediately see the results of their labours. The time consuming nature of construct elicitation (Fransella et al., 2004; Jankowicz, 2004), combined with ambiguity in the respondent's eyes about how repertory test results would subsequently be used in wider planning discussions is a major hurdle to overcome. Canning and Holmes (2007) have suggested that repertory grids offered the Sheffield Galleries and Museums Trust a snapshot of the breadth of respondent opinions regarding the proposed New Deal for Community's Project, as well as a neat visual representation of those opinions for inclusion in published reports. To ensure that such results are usefully employed, community practitioners must be accomplished not only in the administration of the repertory grid test, but also with the analysis of results. Communities must also, from our experience see value in qualitative data as a mechanism for influencing public opinion and debate. We are reminded of a phone conversation that we once had with a community member campaigning for an end to coal seam gas in the Hunter Valley Region of New South Wales (Australia). After describing the merits of the PCT and repertory grid approach, we were told in no uncertain terms that numbers talk. Anything less than 5000 people completing the survey/ interviews would mean that governments and other policy makers would likely put little stock in the validity of the results.

To make repertory grids relevant to community practitioners will, we suggest, require the development of an associated toolkit, one that can be employed by practitioners selectively to affect particular operational objectives. A search for existing scholarship on 'repertory grid toolkits' yielded one series of studies, which have concerned themselves with the way children measure fun (Read & MacFarlane, 2006; Read, MacFarlane, & Casey, 2001; Read, MacFarlane, & Casey, 2002). One of a suite of tools that were developed as part of these studies was the Fun-Sorter. Identified as being loosely based on Kelly's repertory grid the premise and method of employment of the Fun-Sorter was described as follows:

> The Fun Sorter (named after the commonly found children's toy known as the Shape Sorter) is used to compare a set of related [elements] … Loosely based on a repertory grid … the Fun Sorter is made up of n + 1 columns (where n is the number of items being compared), and m + 1 rows (where m is the number of constructs being used). To complete the Fun Sorter, the children need to interpret the construct and then either write a description of the [element] … in blank spaces, or for those children with poor reading or writing abilities, place pre-prepared picture cards on an empty grid. Having ranked the [elements] or placed the cards in this way, a ranked score can be applied to each item/construct under consideration. (Read, 2008, p. 122)

In reporting on the success of the Fun-Sorter as a mechanism for eliciting children's fun constructs; Read (2008) make a number of observations on the adaptability and usability of the tool. Firstly, it was identified that children's propensity to take things literally can cause problems with the use of construct elicitation techniques. Fransella et al. (2004) have defined that a construct is a bipolar dimension, which means that it is distinguished totally from a construct; 'by stating that something is a tree, we are also specifically stating what a tree is not – for instance it is not a bush. The opposite of the concept tree is everything that it not a tree' (Fransella et al. 2004, p. 16). Complexity over how to interpret constructs can similarly cause problems for community members in heterogeneous tourism destination environments on account of the fact that perception of land use is usually not black and white. For instance a repertory grid respondent in Schweinsberg et al. (2012) is identified as attaching the bi-polar descriptor 'environmentally destructive' to the 'pine plantation' land use element.[2] On further examination it was identified that this descriptor had in a sense been forced on the respondent by the triad nature of the repertory grid questioning. Pine plantations were seen as destructive but only in the context of other elements being considered. To limit the effect of this issue Wearing et al. (2014) experimented with the use of written instructions for respondents to aid tourist understanding of the construct elicitation process. Whilst from the researchers' perspective the instructions presented a clear sequential explanation of the steps in the repertory grid process (element selection, construct formulation, element ratings etc.) the abstract nature of the examples, which was a necessary requirement to avoid leading the respondents on the actual subject matter of the survey subsequently created challenges in respondent retention owing to the explanation time involved.

Another lesson identified by Read (2008) in relation to the Fun-Sorter related to the manner in which triad combinations of elements are presented to respondents. The first stage in this process was the selection of elements. As exemplified in Waitt et al. (2003) and Naoia et al. (2006); photographs often serve as surrogates for narrative descriptions of the topics under investigation in environmental management based repertory grid studies. Even when researchers screen the photographs to ensure they were within the respondent's range of convenience through pilot testing and sourcing photographs from local databases of tourism authorities and other regulatory agencies, there is still the potential for confusion. Silverman (2007) has written that photographs allow us a lens on the remarkable in the mundane. How we interpret photos is, however, subjective. The photographer Michal Chelbin has 'warned against the danger of trying to construct a definitive account of her image … I try to create a scene where there is a mixture of straight information and riddles' (Silverman, 2007, p. 5). The effect of this ambiguity is that much as Read (2008) noted that children need to understand what picture cards (as elements) represent; so to must rural residents often be equated with the purpose behind each element chosen for the study. In Schweinsberg et al. (2012) evidence was observed of repertory grid respondents debating aspects of the composition of an element image in ways that were irrelevant to the formation of their construct or the purpose of the study. For example, is a photo of a family camping in a national park to be taken as a representation of an ecotourism experience or family interaction in a holiday period? Whilst the distinction is subtle, it is often amplified in circumstances where respondents may have an ideological objection to the very premise of a land use being portrayed.

Returning to wicked problems, Brooks and Champ (2006) note that to engage with wicked problems we must seek to inclusively engage with all stakeholders and promote harmony and discussion amongst those with disparate viewpoints. Repertory Grid Analysis certainly has a role to play in such debates, on account of its ability through construct formation processes to shine a light on individual value positions. The potential for disagreement over the premise of the chosen elements presents a real challenge in the development of an adaptable repertory grid tool. Further complicating the task is the time intensiveness of the administration of the repertory grid instrument, which Pike (2007) has attempted to circumvent by administering repertory grids in a group setting. Such an approach is within the rules of established PCT pedagogy, but is characterized by a range of challenges including a limited ability to clarify any ambiguity in a respondent's mind over the purpose of an image, or in the meaning behind a construct. Wearing et al. (2014) have attempted to address this issue by including follow-up conversations after each construct is elicited, but this only adds to the time consuming nature of the repertory grid elicitation exercise.

Conclusion

Personal construct theory and repertory grids tests are a theoretical and methodological approach with a long history of successful application in the area of tourism studies. Their ability to be applied in individual and group contexts, combined with their ability to shed light on the subjective scripts of stakeholders in contested terrains marks them out as an important complimentary technique for those interested in consultative ecotourism planning. Drawing from our own work in the repertory grid space and other published scholarship we have in the present paper sought to outline some of the challenges that would be faced in the development of repertory grid instruments that can be employed by local people themselves in contested rural ecotourism locales. Brooks and Champ (2006, p. 795) have suggested that to overcome wickedness requires that 'decision makers allow themselves to be directly informed by local positions and knowledge'. We hope that the ideas contained herein will continue the discussion of how such outcomes can be achieved.

Notes

1. For readers interested in step by step instructions for grid formation, construct elicitation and analysis we would direct them to Jankowicz (2004) *The Easy Guide to Repertory Grids.*
2. In this study the repertory grid respondent was asked to interpret (or construe) the relationship between a set of six land use elements commonly found in the forests of the Eden region of the NSW Far South Coast. These elements included primary industry elements (pine plantations, wood chipping) and nature based tourism elements (bushwalking and 4x wheel driving).

Disclosure statement

No potential conflict of interest was reported by the authors.

References

Bell, R. (1997). *Using SPSS to analyse repertory grid data.* University of Melbourne (Department of Psychology).

Bello, F. G., Lovelock, B., & Carr, N. (2017). Constraints of community participation in protected area-based tourism planning: The case of Malawi. *Journal of Ecotourism, 16*(2), 131–151.

Botterill, T., & Crompton, J. (1987). Personal constructions of holiday snapshots. *Annals of Tourism Research, 14*(152–156).

Botterill, T., & Crompton, J. (1996). Two case studies exploring the nature of the tourist's experience. *Journal of Leisure Research, 28*(1), 57–83.

Brooks, J. J., & Champ, P. A. (2006). Understanding the wicked nature of "unmanaged recreation" in Colorado's Front Range. *Environmental Management, 38*(5), 784–798.

Camillus, J. C. (2008). Strategy as a wicked problem. *Harvard Business Review, 86*(5), 98.

Canning, C., & Holmes, K. (2007). Community consultation in developing museum projects: A case study using the repertory grid technique. *Cultural Trends, 15*(4), 275–297.

Cater, E. (2006). Ecotourism as a western construct. *Journal of Ecotourism, 5*(1–2), 23–39.

Ceballos-Lascurain, H. (1987). The future of ecotourism. *Mexico Journal, 1*(17), 13–19.

Chang, R. C., & Mak, A. H. (2018). Understanding gastronomic image from tourists' perspective: A repertory grid approach. *Tourism Management, 68*, 89–100.

Christie, M. J., Rowe, P. A., & Pickernell, D. (2009). Unpacking a wicked problem: Enablers/impediments to regional engagement. *Australian Journal of Public Administration, 68*(1), 83–96.

Churchman, C. (1967). Wicked problems. *Management Science, 14*(4), B141–B146.

Coakes, S., Fenton, M., & Gabriel, M. (1999). Application of repertory grid analysis in assessing community sensitivity to change in the forest sector. *Impact Assessment and Project Appraisal, 17*(3), 193–202.

Cobbinah, P. B. (2015). Contextualising the meaning of ecotourism. *Tourism Management Perspectives, 16*, 179–189.

Coombs, W. T., & Holladay, S. J. (2018). Social issue qua wicked problems: The role of strategic communication in social issues management. *Journal of Communication Management, 22*(1), 79–95.

Coshall, J. (1991). An appropriate method for eliciting construct subsystems from repertory grids. *The Psychologist: Bulletin of the British Psychological Society, 4*, 354–357.

Coshall, J. (2000). Measurement of tourists' images: The repertory grid approach. *Journal of Travel Research, 39*, 85–89.

Donohoe, H. M., & Needham, R. D. (2006). Ecotourism: The evolving contemporary definition. *Journal of Ecotourism, 5*(3), 192–210.

Duffy, L. N., Kline, C., Swanson, J. R., Best, M., & McKinnon, H. (2017). Community development through agroecotourism in Cuba: An application of the community capitals framework. *Journal of Ecotourism, 16*(3), 203–221.

Elden, S. (2004). *Understanding Henri Lefebvre: A&C Black*. London: Continuum.

Embacher, J., & Buttle, F. (1989). A repertory grid analysis of Australia's image as a summer destination. *Journal of Travel Research, 27*, 3–7.

Fransella, F., Bell, R., & Bannister, D. (2004). *A manual for repertory grid technique* (2nd ed.). London: Academic Press - A Subsidiary of Harcourt Brace Jovanic Publishers.

Garrod, B. (2003). Local participation in the planning and management of ecotourism: A revised model approach. *Journal of Ecotourism, 2*(1), 33–53.

Halfacree, K. (2007). Trial by space for a 'radical rural': Introducing alternative localities, representations and lives. *Journal of Rural Studies, 23*(2), 125–141.

Hall, C., Gössling, S., & Scott, D. (2015). Tourism and sustainability: An introduction. In C. M. Hall, S. Gössling, & D. Scott (Eds.), *The Routledge handbook of tourism and sustainability* (pp. 1–12). London: Routledge.

Hankinson, G. (2004a). The brand images of tourism destinations: A study of the saliency of organic images. *Journal of Product & Brand Management, 13*(1), 6–14.

Hankinson, G. (2004b). Repertory grid analysis: An application to the measurement of destination images. *International Journal of Nonprofit and Voluntary Sector Marketing, 9*(2), 145–153.

Head, B. W. (2008). Wicked problems in public policy. *Public Policy, 3*(2), 101.

Home, R., Bauer, N., & Hunziker, M. (2007). Constructing urban green spaces: An application of Kelly's repertory grid. *Tourism Review International, 62*(3+4), 47–52.

Jankowicz, D. (2004). *The easy guide to repertory grids*. Chichester: John Wiley and Sons.

Jenkins, O. (1999). Understanding and measuring tourist destination images. *International Journal of Tourism Research, 1*(1), 1–15.

Jentoft, S., & Chuenpagdee, R. (2009). Fisheries and coastal governance as a wicked problem. *Marine Policy, 33*(4), 553–560.

Johns, G. (2008). The Northern Territory Intervention in Aboriginal affairs: Wicked problem or wicked policy? *Agenda: A Journal of Policy Analysis and Reform, 15*(2), 65–84.

Jones, S. (2005). Community-based ecotourism: The significance of social capital. *Annals of Tourism Research, 32*(2), 303–324.

Kawaf, F., & Tagg, S. (2017). The construction of online shopping experience: A repertory grid approach. *Computers in Human Behavior, 72*, 222–232.

Kelly, G. (1963). *A theory of personality: The psychology of personal constructs*. New York: The Norton Library W.W. Norton and Company Inc.

Kelly, G. (1969). Hostility. *Clinical psychology and personality: The selected papers of George Kelly*. New York: John Wiley & Sons.

Khan, A. S., & Neis, B. (2010). The rebuilding imperative in fisheries: Clumsy solutions for a wicked problem? *Progress in Oceanography, 87*(1), 347–356.

Lawton, L. J. (2005). Resident perceptions of tourist attractions on the Gold Coast of Australia. *Journal of Travel Research, 44*(2), 188–200.

Lazarus, R. J. (2008). Super wicked problems and climate change: Restraining the present to liberate the future. *Cornell Law Review, 94*, 1153.

Lefebvre, H. (1991 [1974]). *The production of space* (Vol. 142). Oxford: Blackwell.

Levin, K., Cashore, B., Bernstein, S., & Auld, G. (2012). Overcoming the tragedy of super wicked problems: Constraining our future selves to ameliorate global climate change. *Policy Sciences, 45*(2), 123–152.

Liu, J., Qu, H., Huang, D., Chen, G., Yue, X., Zhao, X., & Liang, Z. (2014). The role of social capital in encouraging residents' pro-environmental behaviors in community-based ecotourism. *Tourism Management, 41*, 190–201.

Mak, A. H., Lumbers, M., Eves, A., & Chang, R. C. (2013). An application of the repertory grid method and generalised Procrustes analysis to investigate the motivational factors of tourist food consumption. *International Journal of Hospitality Management, 35*, 327–338.

Mansfield, Y., & Ginosar, O. (1994). Evaluation of the repertory grid method in studies of locals' attitude towards tourism development processes. *Environment and Planning A, 26*, 957–972.

McCool, S. F., Patterson, M. E., W., & D. (2000). Trends in recreation, tourism and protected area planning. In W. Gartner & D. Lime (Eds.), *Trends in Outdoor Recreation, Leisure and Tourism* (pp. 111–119). Oxon: CABI.

McWilliams, S. A. (2008). Ontological acceleration and human agency. *Personal Construct Theory and Practice, 5*, 60–67.

Naoia, T., Aireyb, D., Iijimaa, S., & Niininenc, O. (2006). Visitors' evaluation of an historical district: Repertory grid analysis and laddering analysis with photographs. *Tourism Management, 27* (3), 420–436.

Niblo, D., & Jackson, M. (2004). Model for combining the qualitative emic approach with the quantitative derived etic approach. *Australian Psychologist, 39*(2), 127–133.

Odeh, M. (2013). The wicked problem of tourism: Economic benefits versus socio-cultural impacts. In P. Diaz (Ed.), *Tourism as a Tool for Development* (pp. 123–135). Southhampton WIT Press: WIT Press.

O'Doherty, D., De Cock, C., Rehn, A., & Ashcraft, K. L. (2013). New sites/sights: Exploring the white spaces of organization. *Organization Studies, 34*(10), 1427–1444.

Pearce, P. (2013). *The social psychology of tourist behaviour: International series in experimental social psychology* (Vol. 3). Amsterdam: Elsevier.

Pearce, P., & Packer, J. (2013). Minds on the move: New links from psychology to tourism. *Annals of Tourism Research.*

Pearce, J., & Pearce, P. (2017). Retro-methodologies: Insights for city tourism research. *International Journal of Tourism Cities, 3*(1), 17–29.

Pearce, P., & Stringer, P. (1991). Psychology and tourism. *Annals of Tourism Research, 18*(1), 136–154.

Perec, G. (1997). *Species of spaces and other pieces*. London: Penguin.

Pike, S. (2007). Current issues in method and practice repertory grid analysis in group settings to elicit salient destination image attributes. *Current Issues in Tourism, 10*(4), 378–392.
Pike, S. (2012). Destination positioning opportunities using personal values: Elicited through the repertory test with laddering analysis. *Tourism Management, 33*(1), 100–107.
Pike, S. (2018). Personal construct theory and the repertory test. In R. Nunkoo (Ed.), *Handbook of research methods for tourism and hospitality management* (pp. 222–230). Abingdon: Edward Elgar Publishing.
Pike, S., & Kotsi, F. (2016). Stopover destination image—using the repertory test to identify salient attributes. *Tourism Management Perspectives, 18*, 68–73.
Pike, S., & Ryan, C. (2004). Destination positioning analysis through a comparison of cognitive, affective, and conative perceptions. *Journal of Travel Research, 42*(4), 333–342.
Potter, R., & Coshall, J. (1984). The hand anlaysis of repertory grids: An appropriate mehod for third world environmental studies. *Area, 16*, 315–322.
Read, J. (2008). Validating the Fun Toolkit: An instrument for measuring children's opinions of technology. *Cognition, Technology & Work, 10*(2), 119–128.
Read, J., & MacFarlane, S. (2006). *Using the fun toolkit and other survey methods to gather opinions in child computer interaction*. Paper presented at the proceedings of the 2006 conference on interaction design and children.
Read, J., MacFarlane, S. J., & Casey, C. (2001). Expectations and endurability-measuring fun. *Computers and fun, 4*.
Read, J., MacFarlane, S., & Casey, C. (2002). *Endurability, engagement and expectations: Measuring children's fun*. Paper presented at the interaction design and children.
Reggers, A., Grabowski, S., Wearing, S., Chatterton, P., & Schweinsberg, S. (2016). Exploring outcomes of community-based tourism on the Kokoda Track, Papua New Guinea: A longitudinal study of participatory rural appraisal techniques. *Journal of Sustainable Tourism, 24*(8–9), 1139–1155.
Rittel, H. W., & Webber, M. M. (1973). Dilemmas in a general theory of planning. *Policy Sciences, 4*(2), 155–169.
Scherl, L. M. (1988). Constructions of a wilderness experience: Using the repertory grid technique in a natural setting. *Australian Psychologist, 23*(2), 225–242.
Scherrer, P., & Doohan, K. (2014). Taming wicked problems: Towards a resolution of tourism access to traditional owner lands in the Kimberley region, Australia. *Journal of Sustainable Tourism, 22*(7), 1003–1022.
Schweinsberg, S., Darcy, S., & Cheng, M. (2017). The agenda setting power of news media in framing the future role of tourism in protected areas. *Tourism Management, 62*, 241–252.
Schweinsberg, S., Wearing, S., & Darcy, S. (2012). Understanding communities' views of nature in rural industry renewal: The transition from forestry to nature-based tourism in Eden, Australia. *Journal of Sustainable Tourism, 22*(2), 195–213.
Schweinsberg, S., Wearing, S., & Wearing, M. (2015). Transforming nature's value – cultural change comes from below: Rural communities, the 'othered' and host capacity building. In Y. Reisinger (Ed.), *Transforming tourism: Host perspectives* (pp. 102–113). Oxfordshire: CABI.
Shultis, J. D., & Way, P. A. (2006). Changing conceptions of protected areas and conservation: Linking conservation, ecological integrity and tourism management. *Journal of Sustainable Tourism, 14*(3), 223–237.
Silverman, D. (2007). *A very short, fairly interesting and reasonably cheap book about qualitative research*. Los Angeles: Taylor & Francis.
Stewart, V. (2004). Business applications of repertory grid chapter 6. Analysis of Grid Data *Enquire Within*.
Stone, M. (2015). Community-based ecotourism: A collaborative partnerships perspective. *Journal of Ecotourism, 14*(2–3), 166–184.
Stringer, P. F., & Pearce, P. L. (1984). Toward a symbiosis of social psychology and tourism studies. *Annals of Tourism Research, 11*(1), 5–17.
Stronza, A., & Gordillo, J. (2008). Community views of ecotourism. *Annals of Tourism Research, 35* (2), 448–468.

Thompson, P. B., & Whyte, K. P. (2012). What happens to environmental philosophy in a wicked world? *Journal of Agricultural and Environmental Ethics*, *25*(4), 485–498.

Tribe, J. (1997). The indiscipline of tourism. *Annals of Tourism Research*, *24*(3), 638–657.

Waitt, G., Lane, R., & Head, L. (2003). The boundaries of nature tourism. *Annals of Tourism Research*, *30*(3), 523–545.

Walmsley, D., & Jenkins, J. (1993). Appraisive images of tourist areas: Applications of personal constructs. *Australian Geographer*, *24*(2), 1–13.

Wearing, S., Schweinsberg, S., Lai, P.-H., & Lyons, K. (2014). A discussion of coal seam gas in Australia's hunter valley wine tourism region. *Australasian Parks and Leisure*, *17*(2), 49.

Wearing, S., Schweinsberg, S., & Tower, J. (2016). *The marketing of national parks for sustainable tourism*. Clevedon: Channel View.

Weaver, D. (2005). Comprehensive and minimalist dimensions of ecotourism. *Annals of Tourism Research*, *32*(2), 439–455.

Yeh, D.-Y., & Cheng, C.-H. (2015). Recommendation system for popular tourist attractions in Taiwan using Delphi panel and repertory grid techniques. *Tourism Management*, *46*, 164–176.

Young, M. (1999). The social construction of tourist places. *Australian Geographer*, *30*(3), 373–389.

Community-based ecotourism: beyond authenticity and the commodification of local people

Nantira Pookhao Sonjai, Robyn Bushell, Mary Hawkins and Russell Staiff

ABSTRACT
Community-based ecotourism (CBET) aims to empower local communities through participation. Thailand is a good example of a developing country where the local community requires long-term support from stakeholders for empowerment. Given that the inauthenticity and commodification of CBET have yet to be discussed in terms of the influences of policy and external stakeholders, this research connects the missing link in the literature between policy, tourism stakeholders and practices in relation to the authenticity and commodification of CBET. In order to gain an understanding of the policy related to the CBET operation, central Thai government agencies and related non-governmental organisations (NGOs) participated in in-depth interviews. This research involved a case study in the Khiriwong community, the first CBET community (since 1996) in Thailand, through the application of ethnographic techniques. The findings reveal that since the notion of CBET relies heavily on aspects of community participation, cohesion, forest guardianship, self-reliance and sustainability, all of which are part of the picturesque ideal of romanticism, the community itself becomes associated with the identity, subject to the national policy and the expectations of external stakeholders, which eventually fall into the commodification of staged authenticity.

Introduction

Globalisation and technological development have fuelled significant changes throughout the world. Globalisation is one of the most powerful factors in the continuing rapid growth of the tourism industry, while technological advancements have facilitated people's ability to travel (Singh, 2012). Described by Harvey (1989) in *the Condition of Postmodernityas a time–space compression that eventually*

Since the 1970s, tourism scholars have discerned numerous unfavourable aspects of the tourism phenomena affecting both societies and natural environments (Buckley, 2012). The term 'sustainable development' (SD) officially coined by the World Commission on Environment and Development (WCED) in the publication, the Brundtland report, *Our Common Future*, in 1987, which promoted the term 'SD' and led the endeavour of environmental reform. The 1992 United Nations Conference on Environment and Development (UNCED), also known as the Rio Earth Summit, resulted in many international agreements such as *Agenda 21*. These in turn led to many nations, including the Association of Southeast Asian Nations (ASEAN) establishing guideline for local communities with an attention to community participation in decision-making about development (UNCED, 1993).

Organisations in the tourism sector, including the World Travel and Tourism Council (WTTC), the World Tourism Organization (WTO) and the Earth Council, collaborated and introduced the principles of *Agenda 21* to the tourism industry, which emphasised on decentralisation, encouraging local authorities to manage sustainability in tourism relating to the conservation of nature and the environment, awareness and education, together with the cooperation of local communities in its development. Due to the movement of tourism stakeholders towards alternative forms of tourism, including ecotourism and the community-based tourism (CBT) approach, have evolved under the umbrella of 'sustainable tourism'.

Thailand is an example of a developing country that has embraced the notion of SD in its policy, and has emphasised greater equilibrium between the economy, environment and society, rather than placing the sole emphasis on the economy, from its 7th National Economic and Social Development Plan (NESDP 1992–1996) onwards. Subsequently, community-based ecotourism (CBET) was recognised because of the emphasis of decentralised management, as highlighted in Plan 8 (1997–2001) (Sarobol, Wongtabtim, & Songpornvanich, 2002).

Both national policy and external stakeholders are influential, empowering local communities' participation in planning, developing and operating tourism businesses. Jamal and Getz (1995, p. 193) indicated that "external forces … also influence the destination's stakeholders. The destination domain is thus characterised by an 'open-system' of interdependent, multiple stakeholders, where the actions of one stakeholder impacts on the rest of the actors in the community … ". Through continuous encounters with external influences over a long period of time, local communities have learnt and adapted to the expectations of tourism demand. The aim of this study was to identify the relationship between the external forces with influence on a local CBET operation and the impacts of commodification and staged authenticity inherent in the processes of CBET.

The Thai development style with regard to quality of life

The Thai style is embedded in hybridity between the major Western development approach and the alternative development approach. According to Aekphachaisawat (2011), the major Western development approach adopted by the Thai nation that was directed towards an evolution of capitalism that emphasised monetary growth was based on modernisation theory, dependency theory, growth theory and globalisation theory, and was constituted by the NESDP Plans 1–7 (1961–1996). By contrast, in Plans 8–11 (1997–2016), the focus has evolved into a transition of human development and the strengthening community through application of the alternative development approach. This approach consists of three socio-economic models, including the Buddhist model, political economy and community culture (*wattanatham chumchon*). These components operate independently as well as cooperatively, and are also embedded in the characteristics of the Sufficiency Economy Philosophy (SEP).

The SEP, *setthakit pho phiang* - or the 'just-enough economy', is a philosophy developed by His Majesty King Bhumibol Adulyadej (Rama IX) and was featured in the Thai development scheme. The SEP was first introduced by King Rama IX at a graduation ceremony at Kasetsart University, Bangkok, Thailand in 1974. It was re-emphasised during the Asian economic crisis of 1997–1998, to help Thais to prepare for distress caused by an unfavourable internal and external economic crisis, urging Thai people to put this concept into practice.

The philosophy has been integrated into the national agenda. Originating in Plan 9–11 (2002–2016) it is again highlighted in the current Plan 12 (2017–2021), as a vital element of the national development strategy. In practice, the concept has been accelerated at the provincial and district level by both government-sector and non-governmental organisations (NGOs). The SEP has been merged into the school curriculum to strengthen the concept through the next generation, while Thai national higher government leaders are obligated to be educated about the SEP via the Office of the Civil Service Commission, under the 'Toward High Potential Leaders' programme. Any academic research project associated with the SEP, community self-reliance or sustainability conspicuously receives enhanced support in the form of government grants (see for example Seubsman, Kelly, & Sleigh, 2013). In similar fashion, the Petroleum Authority of Thailand (PTT) Public Company Limited has espoused the concept of the SEP within local communities as part of its corporate social responsibility campaign, which aims to develop 84 subdistricts (*Tambol*) across the country, working towards self-reliant communities and sustainability. Apart from endorsement by government agencies, the concept has been extensively adopted, including by NGOs and tourism stakeholders, since it promotes the well-being of people at the grassroots.

Despite some scholarly questions about the notion of the SEP (see Ivarsson, 2007; Ivarsson & Isager, 2010), many research projects have been conducted, resulting in many books and articles written by Thai researchers and academics that endorse the SEP as a panacea, and illustrate its successful application towards a 'better quality of life' for the nation's people (see Grossman & Faulder, 2011; Kantabutra, 2006; Mongsawad, 2008, 2012; Piboolsravut, 2004; Pruetipibultham, 2010; Sathirathai & Piboolsravut, 2004; Savetpanuvong, Tanlamai, & Lursinsap, 2011; Traithip, Kohtbantau, & Baojai, 2008). The SEP is guiding Thailand towards balance in its development scheme and SD in the globalised world (Piboolsravut, 2004), and aims to drive localisation for a sustainable community (as a national priority), as a bridge between localisation and globalisation (Lohchaiyakul, 2012).

The concepts of SD and SEP are consistent (Mongsawad, 2012; Pruetipibultham, 2010) in that both aim to enhance the quality of human life, particularly people's well-being. Daniels (2003) has suggested the ideal of SD derives from a fusion of East and West. It is development that 'meets the needs of the present without compromising the ability of future generations to meet their own needs' (WCED, 1987, p. 43). In the same vein, the SEP begins with the strengthening of the economic foundation by ensuring that the majority of the population have enough to live on before pursuing a more advanced level of economic development by selling the operational surplus to the outside world (Grossman & Faulder, 2011). It does not reject economic theory, economic progress and globalisation. Accordingly, both concepts accept that globalisation unavoidably originates with development, but that global development with local economic viability, sociocultural equitability and environmental integrity is essential, and that community strength and self-reliance (or empowerment in the context of CBET) through local participation can be applied to improve human well-being as a development goal.

Ecotourism, the community-based approach and the application of the SEP

In line with the SEP and SD, sustainable tourism aims to enhance the quality of life of both hosts and guests, and principally focuses on both the minimisation of negative impacts and the promotion of the benefits for the destination for future generations, while satisfying the needs of both local hosts and visitors (Jules, 2005; United Nations Environment Programme & United Nations World Tourism Organization, 2005).

Ecotourism, as a subset of sustainable tourism has been categorised as an alternative form of tourism (Diamantis 1998; cited in Diamantis & Ladkin, 1999; Weaver, 2001). In Thailand, alternative tourism (AT) was practiced in the 1980s by applying the term *kanthongthiao choeng anurak* ('conservation tourism') and *kanthongthiao choeng nivet* ('ecological tourism') in 1994 and 1998, respectively (Leksakundilok, 2004b). As environmental concerns noted in Plan 7 (1992–1996), ecotourism was officially embraced. In 1997, ecotourism was proclaimed as part of the national tourism master plan (Kontogeorgopoulos, 1999) and in 2001 the *National Ecotourism Action Plan* 2002–2006 was launched. Various plans and projects about ecotourism were subsequently launched with support from government and the Ecotourism Cooperative Network, the Thai Ecotourism and Adventure Travel Association and the Foundation for the Protection of Environment and Tourism, were all formed.

The shift in thinking can be observed from Plan 6 (1987–1991) which considered tourism as a source of economic generation, whereas the commencement of human development in Plan 8 (1997–2001) was a transformational phenomenon in Thai society in relation to the strengthening community stream and sustainable tourism, including the CBET realm (Sarobol et al., 2002). As a consequence of Plan 9, the SEP has likewise been applied to tourism. The Ministry of Tourism and Sports (MOTS) has infused the philosophy in the formulation of latest national tourism policies (MOTS, 2012). Similarly, the Tourism Authority of Thailand (TAT) under Suraphon Svetasreni, a former governor, has claimed that the concept of *balance* or a *middle path* has been employed in various marketing efforts, such as in the *Travel with a New Heart for a Sustainable Thailand* campaign in 2011 (TATnews, 2011). Beyond this endorsement by government organisations, the concept of the SEP is also being applied and embedded in the practice of the Thai tourism sector (see Chartrungruang, 2011; Theerapappisit, 2003, 2009) and in CBT (see Chemnasiri, 2013; Peerapun, 2012; Sangragsa & Lukhananuluk, 2013). With regard to product positioning, MOTS is heavily promoting CBT and local communities under various campaigns such as 'Discover Thainess 2015'and 'Amazing Thailand Go Local' in 2018. With the marketing of stereotypical images, 'Thainess' sells quite well in the tourism industry, and travel to local communities is considered the way to experience the 'real and authentic Thai'. As a consequence, CBT propagates a culturally specific version of the 'real' Thai.

Stakeholder theory: communities in need of external support

Community-based ecotourism aims to integrate conservation and development to work towards sustainability, and is considered as a postmodern development agenda that emphasises small-scale, decentralised, locally oriented development, with a sceptical attitude towards rational planning (Butcher, 2007, p. 28). It utilises a neopopulist approach that engages with the concept of active participation and demonstrates a bottom-up approach to development. The SEP consistently aims to integrate a top-down with a bottom-up development approach (Mongsawad, 2012), to prepare local communities to adapt to change – for example, the impacts of tourism – and to support traditional Thai values of harmony and cooperation (Unger, 2009). In the context of tourism, collaboration among stakeholders in the decision-making process enhances the ability to focus on self-reliance and allows them to enjoy shared ownership (Medeiros de Araujo & Bramwell, 1999). Accordingly, the aim of the SEP is to strengthen human capital in the interests of individual self-reliance and a self-reliant community, respectively, and the goal of the self-reliant community is to achieve sustainability. Conversely, Ryan (2002, p. 17) has suggested that 'in addressing the issues of "sustainability" concepts akin to those of social tourism need to be applied, albeit in a wider context of stakeholder theory'.

Stakeholder theory originates from the relationships between an organisation's stakeholders, and was first mentioned by Freeman (1984). He defined a stakeholder as 'any group or individual who can affect or is affected by the achievement of the organisation's objectives' (p. 46). Freeman pointed out that other external (to the organisation) groups can affect the organisation. In the same vein, when one considers a tourism community as an organisation, external stakeholders have the ability to affect the local community. Apart from its recognition in the organisational sector, stakeholder theory is found in community participation and public involvement related to a democratic perspective. In tourism development moving toward sustainability, stakeholder support and involvement is vital for democratic empowerment which is necessary for tourism operation (Byrd, 2007; Byrd, Bosley, & Dronberger, 2009; Byrd, Cárdenas, & Greenwood, 2008; Graci, 2013; Jamal & Getz, 1995; Medeiros de Araujo & Bramwell, 1999; Murphy, 2013; Timothy, 1998) and stakeholder involvement also has the capacity to reduce potential conflict among participants that might occur in the future of tourism operation (Bramwell & Lane, 1999; Yuksel, Bramwell, & Yuksel, 1999).

The stakeholder approach encompasses both normative and instrumental value: Mannigel (2008) suggested that in the participatory approach, stakeholders should be considered as a means (constantly improving the CBET operation) and an end (empowerment). This means that in order to develop tourism sustainably, local people will continually have encounters with the external stakeholders. Also, local stakeholders should actively participate in tourism planning and management (Gunn 1994; cited in Byrd, 2007; Mannigel, 2008).

Guidance and support from external stakeholders is vital for the CBET development. Fletcher (2009, p. 272) mentioned that 'many elements of ecotourism discourse are likely to be quite alien to members of the rural communities in less-developed and non-Western societies where ecotourism is promoted'. Therefore, for local communities in a developing country, such as Thailand, to be able to collaborate in decision-making, they firstly need capacity building. Frequently, knowledge and ability building passes to the hand of external bodies ranging from government agencies to international and domestic NGOs and academics. In the process of capacity-raising, the knowledge is specifically determined by the knowledge providers in advance and is transferred from the external agencies to the local community in the form of education and training.

However, Fletcher (2009) points out that while ecotourism planners may believe that they are promoting the local economy, the 'education' provided may lead to cultural change at the local destination. For example, new knowledge concerning the human–environment relationship may be alien to local people's perceptions (West & Carrier, 2004). Thus, the relationship among stakeholders in the CBT approach is complex: 'between local–traditional and scientific knowledge, and cultural change' (Jamal & Stronza, 2009). This can weaken local knowledge, wisdom and culture in order to replace it with universal knowledge when there is an encounter between internal and external stakeholders, such as the human–nature relationship with regard to sustainability that anthropologists believe is culturally constructed (see Bushell, 1999; Hutchins, 2007; Ingold, 2000; Staiff, 2008), such as the belief in Buddhist ecology, related to the ordaining of trees in Thailand for conservation purposes (see Falvey, 2005; Forsyth & Walker, 2008). In contrast, with stakeholder collaboration, the voice of local stakeholders is heard, resources are gained from various participants (Jamal & Stronza, 2009) and the local knowledge capabilities of local stakeholders are recognised (Bramwell & Lane, 1999; Yuksel et al., 1999).

Power, commodification and authenticity

According to the highlighted literature on the community-based approach, the participation of the local community in tourism development aims to empower the local community. Empowerment means to increase the ability of a person or group of people (Butcher, 2007). Whilst in theory this is good, the notion of power imbalance has long been a concern. Byrd (2007) proposed that to balance power, every stakeholder should collaborate in making decisions. Nevertheless, to develop CBET in developing countries such as Thailand where local people have limited knowledge and experience with tourism and business development, the collaboration might not bring equality and the stakeholders who have more knowledge and agency tend to direct local people and communities.

In relation to the power imbalance of tourism stakeholders in the CBT approach, Jamal and Getz (1995, p. 193) point out that 'no single organisation or individual can exert direct control over the destination's development process'. According to Foucault (1982), power is dynamic. Correspondingly, power is always exercised, imbalanced, dynamic and complex among various stakeholders throughout the development and operational period. Manwa (2003), applying stakeholder theory in his study, found that only people with power received tourism benefits, at the expense of hidden stakeholders.

The issue of power imbalance likewise influences the way in which the community commoditises its sociocultural uniqueness of people and place, and turns them into commodities for the pleasure of tourists. Despite the community-based approach aiming to shift power into the hands of the local community, in order to control, protect and revitalise the culture, or to decide what to present and sell to the tourists, the external power via the process of stakeholder collaboration continues to influence local sociocultural commodification. This commodification revolves around the notion of authenticity (Cohen, 1988; Greenwood, 1989; MacCannell, 1976). The commodification of tourist space in turn creates 'staged authenticity' (MacCannell, 1973, 1999). The front is the 'placersqseen by guests and the back is the private space preserved for the hosts. MacCannell points out that 'false backs' are means by which tourists are deluded into thinking that they have been permitted to access the authentic. Moreover, he noted that the host community can attach to the image as known by outsiders and perceive itself as an ethnic attraction. Any change in this ethnic identity may cause unfavourable impacts for them, such as lower incomes from tourism. Thus, the community 'is frozen in an image of itself or museumized' (MacCannell, 1984, p. 388). This is seen in the continuation with traditional dress that has otherwise been replaced over time by modern dress, or the wearing of this traditional dress that might normally not be worn any more or used only for special occasions, but becomes the 'uniform' of everyday in tourism spaces.

Methodology

The fieldwork was divided into two rounds. The first round of fieldwork, undertaken over two months, was allocated for interviewing ecotourism stakeholders in order to understand the policy related to the CBET operation. 14 key ecotourism informants, including representatives of government authorities and NGOs, were interviewed. A semi-structured in-depth interview was applied. The past experiences, beliefs, attitudes and reactions of the key informants in relation to their engagement with ecotourism and CBET policy and planning were thoroughly verified in the interviews. The second round of fieldwork was conducted over one and a half months at Khiriwong, a CBET community. Ethnographic techniques were applied to explore the internal dimensions of the CBET community and the relationships between the local people and the external stakeholders.

The Khiriwong community, the foremost CBET community in Thailand, was chosen as an exploratory case study to obtain an overview of the phenomenon, in order to gather sufficient information to build a description of what is 'going on' (Bouma, 2000). The researcher was introduced to the community by both central and regional CBET specialists, who are closely connected with the leaders and the key participants in the communities. A snowball technique was applied. The ethnographic technique employed a multi-method approach to gain different views from various groups of tourism stakeholders associated with the CBET operation. This method is valued as a tool that enables the researcher to reveal and interpret the complex context of social phenomena as well as to understand other people's life aspects, such as the social networks, kinship ties and face-to-face social relations of a clearly defined geographical locality (Blackshaw, 2010). The researcher recognised the mixing of the *emic* and *etic* perspectives (Jennings, 2010) of villagers and visitors by embracing the ethnographic technique. This also enabled the researcher to scrutinise an in-depth exploration of villagers' values, norms and beliefs, and the practices of their daily lives – that is, their cultures – while gathering the richness of the symbolic (to see things), which can be fully described (interpreting meaning from within a culture) and can tell the *truth* about the social world (Blackshaw, 2010; Geertz, 1983; O'Leary, 2004; Silverman & Seale, 2005).

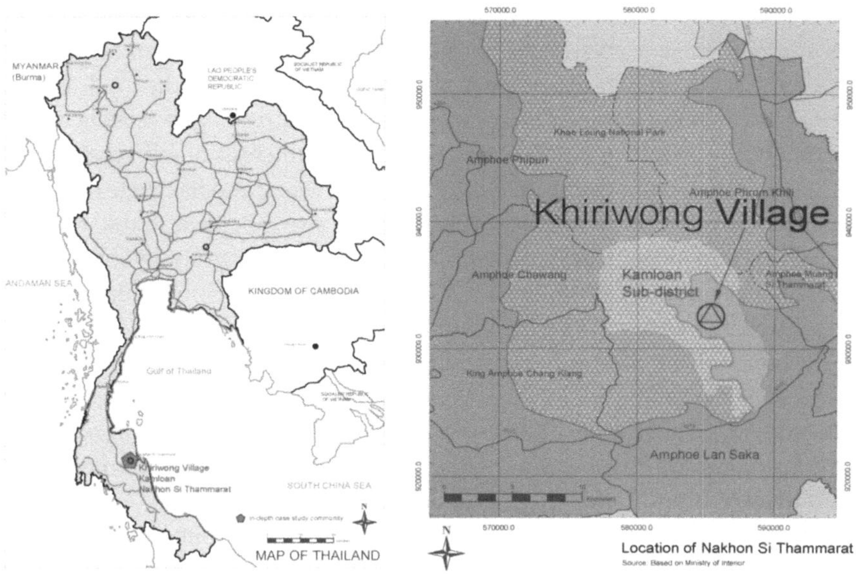

Figure 1. The location of the Khiriwong community. Source: Ministry of Interior and Google Maps.

Case study: the Khiriwong community

Khiriwong is located in Nakhon Si Thammarat province, situated 30 kilometres away from the city. It is believed that the community has been settled here for 300 years since the reign of the King Rama I. The community is located in the foothills of Khao Luang, near the Khao Luang National Park. Khao Luang is covered by tropical evergreen forest and is considered as the highest plateau in the southern part of Thailand (Figure 1).

Khiriwong was once a small community. However, the population has increased to 3120, in 1070 households (Kamloan SAO, 2013), and the community expanded into four nearby villages, Khiriwong (village number, or *Moo*, 5), Khirithong (or *Moo* 8), KhunKhiri (or *Moo* 9) and Khirithum (or *Moo* 10), covering an area of approximately 8173 *rai* in total. The villagers have been making a living by practicing integrated orchard farming of mangosteen, durian, bitter bean, langsat and longkong. The integrated orchard farming provides a profitable return for local people, but, CBET is now seen an alternative activity for them.

Discussion

In 1962, 1975 and 1988, Khiriwong was flooded. Deforestation was believed to be the cause of the disaster. Villagers' lives and property were destroyed. Community collaboration has been used as a tool to move towards self-reliance, to redevelop the community and to rescue local people from poverty resulting from the flood. The kin relationships among the villagers were the reason for this cooperative achievement. Khiriwong has been recognised as a collective community that had defeated the natural disaster of flooding by cooperatively conserving its ecological surroundings, and later on it became known as a noteworthy self-reliant community. This distinctiveness later on turned out to be the community marketing tool for CBET and, correspondingly, to conserve natural resources and avoid further flooding disasters, the villagers agreed that CBET should be a solution in the interests of ecological conservation. The collaboration of Khiriwong can be explained by Unger (2009, p. 143), who has mentioned that 'community elders, the local temple, and resident spirits provided much of the glue necessary for survival and relative harmony'.

Accordingly, the collective action of Khiriwong in the interests of self-reliance has been perceived as a role model within the country, and Khiriwong and its image of CBET have long been associated with this identity. This notion is also described in the work of Lertwicha (1989; cited in in Winichakul, 2008) as Kiriwong is the community that presents the image of self-reliance and anti-capitalism. The achievements in community participation towards self-reliance and empowerment of Khiriwong has been identified by many scholars (see Leksakundilok, 2004a; Leksakundilok & Hirsch, 2008; Opasruttanakorn, Buddharat, & Tongkhundum, 2006; Puan-Auam & Sukphongsri, 2000; Rungklin, 2009; Sukholthajit, 2009; Tungchawal, 2001).Accordingly, the image of Khiriwong in relation to self-reliance, anti-capitalism, community participation and empowerment has been promoted to attract tourists for CBET operation.

Since the notion of the SEP, sustainable tourism and community participation were urged in the national plan, Thai tourism stakeholders have been embracing and implementing the concepts by asserting that, by following these concepts, local people will be directed towards a better quality of life by enhancing ecological integrity, reducing poverty and elevating sociocultural integration and equitability. In cooperation with this promotion, a budget has also been issued to villages all over the country to input how to be self-reliant, and certificates have been issued to those communities that have effectively implemented the concept. Khiriwong was awarded as a community that performed the concept perfectly, and became a role model. These awards include the Best Ecotourism Award from the TAT in 1998, and Regional and National Best Self-sufficient Village in 1886 and 1987. This is similar to Hitchcock's (1999) description of the promotion of tourism that enhances ethno-nationalism. Accordingly, various institutions, particularly study tours organised by government agencies, have visited the community in order to learn about proficiency in self-reliance and ecotourism management.

The CBET operation at Khiriwong was initiated in 1992. The community received external assistance from the state and various organisations, and was supported by several sectors including the Thai Volunteer Services – Responsible Ecological and Social Project (TVS-REST), nowadays known as the Community-based Tourism Institute (CBT-I) and Walailak University funded by the TAT. The Khiriwong Ecotourism Club (KEC) has opened as a centre for community ecotourism management to integrate community activities and alleviate internal and external conflicts among the Khao Luang National Park, the TAT, the KEC and the subdistrict administrative organisation (SAO). The SAO administers the operation of the KEC. The issue of stakeholders is quite complex at Khiriwong, with the ecotourism community situated near the protected area, which is a source of water that supplies the wider area, such that the tourism development affects the ecological diversity, air quality and water quality beyond the community. The complexity of tourism in the protected area impacts multiple stakeholders, ranging from international to the local level (Jamal & Stronza, 2009). This 'stakeholder' approach to conservation is described as the commodification of landscapes as 'natural capital' (Harvey, 2005).

The support tends to decrease the identity of the community since the government agencies offer the community budgets, training, certificates and infrastructure development driven by policy schemes. For example, a government agency initiated a training project to raise handicraft capacity for a cooperative group at Khiriwong. These handicraft skills were also being carried out in other communities over Thailand which might result in similar works rather distinctive local styles. Similarly, local people indicated that academics who visited the community tend to promote their own interests, and are not purely community-focused. For collaboration of external stakeholders, some agencies may develop and manage local communities (Leksakundilok, 2004a). This is consistent with Stevens (2010), who stressed that NGOs, as tourism mediators, may impair local initiatives by promoting their own agendas over those of the host community, and may actually create community segregation rather than collaboration by promoting one subgroup over others (Chernela, 2005). In contrast, ecotourism can be directed towards social improvements if it is operated and managed properly (Johasson & Diamantis, 2004).

The most influential program that added value to tourism at Khiriwong is the One Tambol One Product (OTOP), a project initiated by the government since 2001. OTOP aims to enhance local wisdom by promoting local entrepreneurship towards community-self-reliance. The reputation of OTOP at Khiriwong is one of the reasons for its prominence as a tourist destination. The purchasing of agricultural products as a part of OTOP is a prominent tourist activity here. Many products derived from local wisdom at Khiriwong have received awards from the government, gaining three to five stars under the OTOP scheme. Khiriwong has distinctively received a range of other awards apart from OTOP. With this reputation and these certificates, the community identity has been constructed. Accordingly, the identity has been unconsciously dominated by the authorities, which have acquired power over the residents, hence embedding unequal power relations (Bramwell & Sharman, 1999).

The community also become associated with the identity according to how the media has presented it to outsiders with regard to self-reliance and strong local cooperation, and the villagers have subsequently associated with the identity given by the outsiders. The reputation of the CBET operation has inspired the media to come into the community with the offering of free press releases, television programmes and academic papers, which have become sources for a marketing channel that has led to increased numbers of tourists. The media's representation of Khiriwong to outsiders for marketing purposes also verifies and reinforces the community identity. Advertisements typically highlight the CBET in connection with romanticism (primitive, self-reliance, cohesion and participation), nostalgia (for Thais) or exoticism/real/authentic Thailand (for foreigners), to induce tourists who are attracted by this image and have high expectations.

The fieldwork confirmed that the collaborative action of cooperative groups has shifted to become a business-focused exercise, to serve personal interests. It also reveals that the local people are slightly different from the image represented to the outside world. For example, an cooperation is not always observed in the management and operation of the community homestay, which is monitored by the KEC. Due to the geographical location of the four villages, it is difficult to equitably manage and distribute guests who have expressed special requests to stay in the same area. Therefore, there are approximately 30 households have committed to the KEC, while the remainder are operating independently and have their own website. With this management difficulty and the lack of communication between the KEC and homestay members, the villagers are encountering various conflicts of interest. However, many of key tourism informants believed that the conflict of interests and the lack of cooperation among the villagers can be reconciled, as they have strong blood ties and have known each other all their lives. This shows that the community social relationships drive the collective action and conflict reduction among the internal stakeholders in the CBET operation. This is consistent with research that indicates that local community is typically not static nor homogenous (Bushell 1999) and with findings by Novelli and Tisch-Rottensteiner (2012) with a hill tribe village in the north of Thailand. They also observed that tourists need to receive correct information and adjust their expectations to the current state of development in such villages.

Khiriwong's reputation has created a dilemma of identity between the real and the superposed image, which directs the CBET community towards an approach akin to staged authenticity. In accordance with the reputation and the way in which Khiriwong has been described to outsiders, the village has attached itself to the bestowed image and has 'frozen an image of itself or [become] museumized' (MacCannell, 1984, p. 388). Additionally, the villagers have constructed a form of 'imagined identity' (Anderson, 1983, 1991, 2006). This identity was particularly constructed by the notion of policy embedded in the SEP and sustainable tourism, transmitted to the wider community by the educated tourism stakeholders so that the local community could become self-reliant to resist the modern regime of power related to modernisation and capitalisation.

This research demonstrates that the CBET operation empowers the local community in determining what to show to outsiders and there are still various factors associated with power exercised among stakeholders that influence the way local people perceived and constructed their identity for commodification. With the image portrayed by the outside world, the community needs to present itself as cohesive, collaborative, participative and self-reliant to the visitors, who yearn to be educated about these community strengths. These characteristics can sell and can become the 'performance' that is for consumption by the tourists, but not the reality of their everyday lives – a situation that MacCannell (1973) describes as 'staged authenticity'.

Conclusion

This research articulates the connection between policy, tourism stakeholders and practices in relation to the commodification and authenticity of community in CBET operations. In practice, the power of the policy advocated by external stakeholders, and their active participation, become key factors in developing and operating CBET, as local people typically have limited CBET-related knowledge and competency.

The influence of external stakeholder collaboration in the Thai approach to CBET development creates the uniqueness of the CBET operation within the country. This stakeholder approach challenges tourism development in the country with regard to SD, and initialises the commodification of a picturesque 'Thainess' at the local community level, in order to follow the policy of the state, as it relates to the concept of self-reliance. This permeates the complexity of the CBET operation and influences the development of a staged authenticity of the local in response to the demands of tourism. Despite the controllable power of the villagers, the commodification of their culture and nature impacts their way of life. In the open spaces, the villagers are performing what is expected by the tourists, while they tend to operate differently in the confines of their closed space. This suggests that local people have power to choose what to present to outsiders in order to benefit themselves, and are aware of this transactional power. Moreover, this identity and the agency of cohesion have generated positive results rather than negative results, because this reputation has indirectly forced the villagers to commit to more sustainable conservation activities and collective action.

This research suggests that future research and practice should include the importance of local socio-culture as capital for collective action in the stakeholder approach, and should pay attention to the power structure within the local community and between local people and external stakeholders, in order to understand how power is exercised for effective collaboration and achievement of sustainable tourism. Moreover, tourism related stakeholders should question how to sensitively incorporate the sociocultural past and present of ethnic based attractions into tourism planning, development and operation and how to present the current development stage of the community to outsiders so that the community is not forced into a frozen image of itself.

Acknowledgement

The study was possible by the kind support from Dr. Poll Theerapappisit, Geoffrey D. Palmer, Thai policy makers, NGO officers, local academics and villagers at Khiriwong.

Disclosure statement

No potential conflict of interest was reported by the authors.

References

Aekphachaisawat, T. (2011). *Community study* (2nd ed.). Bangkok: Chulalongkorn University Press.

Anderson, B. R. O. G. (1983). *Imagined communities: Reflections on the origin and spread of nationalism*. London: Verso.

Anderson, B. R. O. G. (1991). *Imagined communities: Reflections on the origin and spread of nationalism* (2nd ed.). London: Verso.

Anderson, B. R. O. G. (2006). *Imagined communities: Reflections on the origin and spread of nationalism* (rev. ed.). London: Verso.

Blackshaw, T. (2010). *Key concepts in community studies*. London: SAGE Publications.

Bouma, G. D. (2000). *The research process* (4th ed.). Melbourne: Oxford University Press.

Bramwell, B., & Lane, B. (1999). Collaboration and partnerships for sustainable tourism. *Journal of Sustainable Tourism*, *7*(3–4), 179–181. doi:10.1080/09669589908667335

Bramwell, B., & Sharman, A. (1999). Collaboration in local tourism policymaking. *Annals of Tourism Research*, *26*(2), 392–415. doi:10.1016/S0160-7383(98)00105-4

Buckley, R. (2012). Sustainable tourism: Research and reality. *Annals of Tourism Research*, *39*(2), 528–546. doi:10.1016/j.annals.2012.02.003

Bushell, R. (1999). Global issues for protected areas and nature-based tourism: Case studies of partnerships in Australia addressing some of these issues. In L. Gündling, H. Korn, & R. Specht (Eds.), *Report of the international expert workshop 'case studies on sustainable tourism and biological diversity'* (pp. 94–114). Isle of Vilm: German Federal Agency for Nature Conservation, 2000.

Butcher, J. (2007). *Ecotourism, NGOs, and development: A critical analysis*. Abingdon: Routledge.

Byrd, E. T. (2007). Stakeholders in sustainable tourism development and their roles: Applying stakeholder theory to sustainable tourism development. *Tourism Review of AIEST – International Association of Scientific Experts in Tourism, 62*(2), 6–13.

Byrd, E. T., Bosley, H. E., & Dronberger, M. G. (2009). Comparisons of stakeholder perceptions of tourism impacts in rural eastern North Carolina. *Tourism Management, 30*(5), 693–703. doi:10.1016/j.tourman.2008.10.021

Byrd, E. T., Cárdenas, D. A., & Greenwood, J. B. (2008). Factors of stakeholder understanding of tourism: The case of eastern North Carolina. *Tourism and Hospitality Research, 8*(3), 192–204. doi:10.1057/thr.2008.21

Chartrungruang, B. (2011). The model of sufficiency economy philosophy: Application for tourism industry in Chiang Mai, Thailand. *International Journal of Asian Tourism Management, 2*(2), 134–144.

Chemnasiri, N. (2013). Community potential development for sustainable agro-tourism in Thailand. *Developing Country Studies, The International Institute for Science, Technology and Education (IISTE), 3*(5), 22–26.

Chernela, J. (2005). The politics of mediation: Local–global interactions in the Central Amazon of Brazil. *American Anthropologist, 107*(4), 620–631. doi:10.2307/3567380

Cohen, E. (1988). Authenticity and commoditization in tourism. *Annals of Tourism Research, 15*(3), 371–386.

Daniels, P. L. (2003). Buddhist economics and the environment: Material flow analysis and the moderation of society's metabolism. *International Journal of Social Economics, 30*(1/2), 8–33. Retrieved from http://park.dnp.go.th/visitor/nationparkshow.php?PTA_CODE=1008

Diamantis, D., & Ladkin, A. (1999). The links between sustainable tourism and ecotourism: A definitional and operational perspective. *The Journal of Tourism Studies, 10*(2), 35–46.

Falvey, J. L. (2005). *Religion and agriculture: Sustainability in Christianity and Buddhism*. Adelaide: Institute for International Development.

Fletcher, R. (2009). Ecotourism discourse: Challenging the stakeholders theory. *Journal of Ecotourism, 8*(3), 269–285. doi:10.1080/14724040902767245

Forsyth, T., & Walker, A. (2008). *Forest guardians, forest destroyers: The politics of environmental knowledge in northern Thailand*. Chiang Mai: Silkworm Books.

Foucault, M. (1982). The subject and power. *Critical Inquiry, 8*(4), 777–795.

Freeman, R. E. (1984). *Strategic management: A stakeholder approach*. Boston, MA: Pitman.

Geertz, C. (1983). *Local knowledge: Further essays in interpretive anthropology*. New York, NY: Basic Books.

Graci, S. (2013). Collaboration and partnership development for sustainable tourism. *Tourism Geographies, 15*(1), 25–42. doi:10.1080/14616688.2012.675513

Greenwood, D. J. (1989). Culture by the pound: An anthropological perspective on tourism as cultural commoditization. In V. L. Smith (Ed.), *Hosts and guests: The anthropology of tourism* (2nd ed., pp. 171–185). Philadelphia, PA: University of Pennsylvania Press.

Grossman, N., & Faulder, D. (Eds.). (2011). *King Bhumibol Adulyadej: A life's work: Thailand's monarchy in perspective*. Singapore: Editions Didier Millet.

Harvey, D. (1989). *The condition of postmodernity: An enquiry into the origins of cultural change*. Oxford: Blackwell.

Harvey, D. (2005). *A brief history of neoliberalism*. Oxford: Oxford University Press.

Hitchcock, M. (1999). Tourism and ethnicity: Situational perspectives. *International Journal of Tourism Research, 1*(1), 17–32.

Hutchins, F. (2007). Footprints in the forest: Ecotourism and altered meanings in Ecuador's upper Amazon. *Journal of Latin American and Caribbean Anthropology, 12*(1), 75–103.

Ingold, T. (2000). *The perception of the environment: Essays on livelihood, dwelling and skill*. London: Routledge.

Ivarsson, S. (2007). King, coup and sufficiency economy: The quest for political legitimacy. *NIAS Nytt, 3*, 23–26.

Ivarsson, S., & Isager, L. (2010). *Saying the unsayable: Monarchy and democracy in Thailand*. Copenhagen: Nordic Institute of Asian Studies (NIAS).
Jamal, T. B., & Getz, D. (1995). Collaboration theory and community tourism planning. *Annals of Tourism Research*, *22*(1), 186–204. doi:10.1016/0160-7383(94)00067-3
Jamal, T., & Stronza, A. (2009). Collaboration theory and tourism practice in protected areas: Stakeholders, structuring and sustainability. *Journal of Sustainable Tourism*, *17*(2), 169–189. doi:10.1080/09669580802495741
Jennings, G. (2010). *Tourism research* (2nd ed.). Milton, QLD: John Wiley & Sons.
Johasson, Y., & Diamantis, D. (2004). Ecotourism in Thailand and Kenya: A private sector perspective. In D. Diamantis (Ed.), *Ecotourism: Management and assessment* (pp. 298–311). London: Thomson.
Jules, S. (2005). *A sustainability assessment of trade and liberalization in tourism-services* (Report). International Institute for Sustainable Development. Retrieved from http://www.iisd.org/tkn/pdf/tkn_tourism_stlucia.pdf
Kantabutra, S. (2006). Relating vision-based leadership to sustainable business performance: A Thai perspective. *Kravis Leadership Institute Leadership Review*, *6*, 37–53.
Kontogeorgopoulos, N. (1999). Sustainable tourism or sustainable development? Financial crisis, ecotourism, and the 'amazing Thailand' campaign. *Current Issues in Tourism*, *2*(4), 316–332. doi:10.1080/13683509908667859
Leksakundilok, A. (2004a). *Community participation in ecotourism development in Thailand* (Doctor of Philosophy thesis). University of Sydney, University of Sydney.
Leksakundilok, A. (2004b). *Ecotourism and community-based ecotourism in the Mekong region* (Vol. 10). Sydney: Australian Mekong Resource Centre.
Leksakundilok, A., & Hirsch, P. (2008). Community-based ecotourism in Thailand. In J. Connell & B. Rugendyke (Eds.), *Tourism at the grassroots: Villagers and visitors in the Asia-pacific* (pp. 214–235). London: Routledge.
Lohchaiyakul, P. (2012). Alternative marketing: When localization is dominated. *TATejournal*. Retrieved from http://www.etatjournal.com/upload/244/8_Localization.pdf
MacCannell, D. (1973). Staged authenticity: Arrangements of social space in tourist settings. *American Journal of Sociology*, *79*(3), 589–603.
MacCannell, D. (1976). *The tourist: A new theory of the leisure class*. Berkeley, CA: University of California Press.
MacCannell, D. (1984). Reconstructed ethnicity tourism and cultural identity in third world communities. *Annals of Tourism Research*, *11*(3), 375–391.
MacCannell, D. (1999). *The tourist: A new theory of the leisure class* (2nd ed.). Berkeley, CA: University of California Press.
Mannigel, E. (2008). Integrating parks and people: How does participation work in protected area management? *Society & Natural Resources*, *21*(6), 498–511. doi:10.1080/08941920701618039
Manwa, H. (2003). Wildlife-based tourism, ecology and sustainability: A tug-of-war among competing interests in Zimbabwe. *Journal of Tourism Studies*, *14*(2), 45–54.
Medeiros de Araujo, L., & Bramwell, B. (1999). Stakeholder assessment and collaborative tourism planning: The case of Brazil's Costa Dourada project. *Journal of Sustainable Tourism*, *7*(3–4), 356–378. doi:10.1080/09669589908667344
Ministry of Tourism and Sports (MOTS). (2012). *Strategy of ministry of tourism and sports 2012–2016*. Bangkok: MOTS. Retrieved from http://www.mots.go.th/ewt_dl_link.php?nid=3942
Mongsawad, P. (2008). Sufficiency economy: A contribution to economic development. *World Academy of Science, Engineering and Technology*, *2*(5), 241–248.
Mongsawad, P. (2012). The philosophy of the sufficiency economy: A contribution of the theory of development. *Asia-Pacific Development Journal*, *17*(1), 123–143.
Murphy, P. E. (2013). *Tourism: A community approach* (2nd ed.). London: Routledge.
Novelli, M., & Tisch-Rottensteiner, A. (2012). Authenticity versus development: Tourism to the hill tribes of Thailand. In O. Moufakkir & P. M. Burns (Eds.), *Controversies in tourism* (pp. 54–72). Wallingford: CAB International.
O'Leary, Z. (2004). *The essential guide to doing research*. London: SAGE Publications.

Opasruttanakorn, O., Buddharat, C., & Tongkhundum, T. (2006). Needs of Khiriwong people to study English for tourism business. *Journal of Yala Rajabhat University*, *1*(2), 112–119.

Peerapun, W. (2012). Participatory planning approach to urban conservation and regeneration in Amphawa community. *Asian Journal of Environment-Behaviour Studies*, *3*(7), 35–44.

Piboolsravut, P. (2004). Sufficiency economy. *ASEAN Economic Bulletin*, *21*(1), 127–134. doi:10.2307/25773806

Pruetipibultham, O. (2010). The sufficiency economy philosophy and strategic HRD: A sustainable development for Thailand. *Human Resource Development International*, *13*(1), 99–110. doi:10.1080/13678861003589073

Puan-Auam, K., & Sukphongsri, S. (2000). *Ecotourism: The study area, the Khiri Wong community, Tambon Kamlon, Amphur Lan Saka, Nakhon Si Thammarat Province*. Retrieved from http://kb.hsri.or.th/dspace/handle/11228/1223

Rungklin, D. (2009). *Consumers behavior and marketing factors: Buying TIE-DYE of Ban Khiriwong's Natural Color TIE-DYE, Community Nakhon Si Thammarat Province* (Master of Business Administration thesis). Walailak University, Nakhon Si Thammarat.

Ryan, C. (2002). Equity, management, power sharing and sustainability – issues of the 'new tourism'. *Tourism Management*, *23*(1), 17–26.

Sangragsa, N., & Lukhananuluk, S. (2013). Research for development and changing in cultural tourism toward creative economy through participation process of sustainable network alliances in Ratchaburi Province. *Journal of Social Sciences, Humanities, and Arts*, *13*(1), 139–156.

Sarobol, S., Wongtabtim, U., & Songpornvanich, S. (2002). *Community-based tourism: Concepts and experiences* [Karntongtiew doei chumchon: naewkind lae prasobkarn]. Chiang Mai: Upper Northern Research Administration Network (UNRN).

Sathirathai, S., & Piboolsravut, P. (2004). *Sufficiency economy and a healthy community*. Paper presented at the 3rd IUCN World Conservation Congress, Bangkok, Thailand. Retrieved from http://pikul.lib.ku.ac.th/FullText_SE/SE010224.pdf

Savetpanuvong, P., Tanlamai, U., & Lursinsap, C. (2011). Sustaining innovation in information technology entrepreneurship with a sufficiency economy philosophy. *International Journal of Innovation Science*, *3*(2), 69–82. doi:10.1260/1757-2223.3.2.69

Seubsman, S., Kelly, M., & Sleigh, A. (2013). The sufficiency economy and community sustainability in rural northeastern Thailand. *Asian Culture and History*, *5*(2), 57–65.

Silverman, D., & Seale, C. (2005). *Doing qualitative research: A practical handbook* (2nd ed.). London: SAGE Publications.

Singh, T. (ed.). (2012). *Critical debates in tourism*. Bristol: Channel View Publications.

Staiff, R. (2008). Cultural inscriptions of nature: Some implications for sustainability, nature-based tourism and national parks. In S. F. McCool & R. N. Moisey (Eds.), *Tourism, recreation, and sustainability: Linking culture and the environment* (2nd ed., pp. 220–235). Wallingford: CAB International.

Stevens, M. (2010). Power disparities and community-based tourism in Vietnam. In S. Gmelch (Ed.), *Tourists and tourism: A reader* (2nd ed., pp. 451–463). Long Grove, IL: Waveland Press.

Subdistrict Administrative Organisation (SAO). (2013). *General information of Kamloan Subdistrict*. Kamloan Subdistrict: Kamloan Subdistrict Administrative Organisation.

Sukholthajit, P. (2009). Khiriwong: A self-reliant community [Ban khunnam Khiriwong: chumchon peungtonaeng]. *Romphruekj*, *27*(1), 93–123.

TATnews Organisation. (2011). ASEAN Tourism Forum 2011 (ATF 2011). *TATnews*. Retrieved from http://www.tatnews.org/press-kits-page/item/43-asean-tourism-forum-2011-atf-2011

Theerapappisit, P. (2003). Mekong tourism development: Capital or social mobilization?. *Tourism Recreation Research*, *28*(1), 47–56.

Theerapappisit, P. (2009). Pro-poor ethnic tourism in the Mekong: A study of three approaches in northern Thailand. *Asia Pacific Journal of Tourism Research*, *14*(2), 201–221. doi:10.1080/10941660902847245

Timothy, D. J. (1998). Cooperative tourism planning in a developing destination. *Journal of Sustainable Tourism*, *6*(1), 52–68. doi:10.1080/09669589808667301

Traithip, T., Kohtbantau, S., & Baojai, P. (2008). The development of the training model for community environmental leaders on forest resource conservation according to sufficiency economy philosophy. *Pakistan Journal of Social Sciences*, *5*(9), 905–908.

Tungchawal, K. (2001). *Sustainable ecotourism in the village of Khiriwong and the Khao Luang National Park, Thailand* (Master of Science thesis). University of Wisconsin-Stout.

UNCED. (1993). *Agenda 21: Programme for action for sustainable development, Rio declaration on environment and development*. Geneva: United Nations Conference on Environment and Development.

Unger, D. (2009). Sufficiency economy and the bourgeois virtues. *Asian Affairs: An American Review*, *36*(3), 139–156. doi:10.1080/00927670903259897

United Nations Environment Programme (UNEP) and United Nations World Tourism Organization (UNWTO). (2005). *Making tourism more sustainable: A guide for policy makers* (Report). United Nations Environment Programme and World Tourism Organization. Retrieved from http://www.unep.fr/shared/publications/pdf/DTIx0592xPA-TourismPolicyEN.pdf

Weaver, D. B. (2001). *Ecotourism*. Milton, QLD: John Wiley & Sons.

West, P., & Carrier, J. G. (2004). Ecotourism and authenticity. *Current Anthropology*, *45*(4), 483–498.

Winichakul, T. (2008). Nationalism and the radical intelligentsia in Thailand. *Third World Quarterly*, *29*(3), 575–591. doi:10.1080/01436590801931520

World Commission for Environment and Development (WCED). (1987). *Our common future*. Oxford: Oxford University Press.

Yuksel, F., Bramwell, B., & Yuksel, A. (1999). Stakeholder interviews and tourism planning at Pamukkale, Turkey. *Tourism Management*, *20*(3), 351–360. doi:10.1016/S0261-5177(98)00117-4

Surfing tourism and local stakeholder collaboration

Nick Towner

ABSTRACT
Sustainable destination development is significantly facilitated by the involvement of all relevant stakeholder groups including the local community. Stakeholder collaboration is crucial for effective tourism development because it has the potential to achieve common goals among all stakeholders involved, but it presents multifaceted challenges for destination management. This paper examines the current stakeholder collaboration in the Mentawai Islands surfing tourism industry by exploring multiple stakeholder perspectives. Field research was conducted in the period between July and September of 2010, through in-depth interviews with local stakeholder representatives. The research indicates that despite an apparent enthusiasm to participate and collaborate in tourism planning, collaboration between stakeholder groups was in fact limited or largely avoided. Corruption, traditional mistrust of government and lack of an organised collaboration initiative are seen as the major barriers to effective collaboration in the Mentawai Islands surfing tourism industry.

Introduction

Collaboration in tourism development should include: multiple stakeholders, both private/public and community/private; cross-sector planning; collective decision-making; and the bridging of cultural differences (Okazaki, 2008, p. 511). Collaboration is a vital component in the tourism-development process, with effective communication being crucial in resolving disputes between stakeholder groups (Gray, 1985, 1989). Okazaki (2008, p. 514) comments that independent attempts by stakeholders to find solutions to their challenges often end in failure as different groups have their own objectives. Solutions to community-based problems are often found through collaboration – solutions are more likely to be found when stakeholders act together than when they act alone. Collaboration also relieves tension between the community and private sector, permitting all stakeholders to be involved in the tourism decision-making process (Jamal & Getz, 1999).

Surf tourism is a relatively recent activity in the Mentawai Islands, Indonesia, and it is a rapidly growing industry, with the number of surf resorts and charter boats increasing every year. The economic benefits that surfing tourism brings to destinations around the world have proven significant (Gough, 1999; Lazarow, 2007; Lazarow, Miller, &

Blackwell, 2008; Martin & Assenov, 2012; Nelsen, Pendleton, & Vaughn, 2007; Ntloko & Swart, 2008). However, in some locations the amount of income that the local community receives has been shown to be limited (Ponting, 2007, 2009; Ponting, McDonald, & Wearing, 2005). Furthermore, indications from two studies, in Costa Rica and Indonesia, suggest surfing tourism at a local scale can be responsible for adverse environmental, socio-cultural and economic impacts (Lazarow, Miller, & Blackwell, 2007; Ponting, 2008; Ponting & O'Brien, 2013). Therefore, constructive collaboration between relevant stakeholders of the Mentawai surfing tourism industry is essential for creating an effective sustainable strategy for future tourism development of the region.

Literature review

Stakeholder theory and sustainable tourism development

Freeman (1984, p. 46) defines a stakeholder as 'any group or individual who can affect or is affected by the achievement of the organisation's objectives.' Donaldson and Preston (1995) offer a narrower and more refined definition, stating that to be identified as a stakeholder, the group or individual must have a legitimate interest in the organisation concerned. Hardy (2001, p. 75) notes both definitions have deficiencies when applied to the context of sustainable tourism, as they place an emphasis on recognition and do not address the concept of feedback. Stakeholders do not just relate to organisations; for example, Steelman's (2001) study of national forest planning in the United States concluded that every person who felt involved in the park has the right to be involved no matter their level of power.

An understanding of local stakeholder perspectives towards tourism and related developments is vital if the industry is going to be managed in a sustainable fashion. It is important to monitor and measure stakeholder attitudes where possible (Stewart & Draper, 2007, p. 7). Burns (2003, p. 38) concludes that tourism built up through embracing a multiple stakeholder approach has a greater propensity for long-term stability because of decreased antagonism between sectors and the fostering of cooperative planning. Conflicts can occur when different stakeholder groups perceive tourism costs and benefits in alternative ways. Therefore, to effectively reduce inherent conflict and increase the sustainability of the tourism development, it is essential that the attitudes and perspectives of stakeholders are identified and understood (Reid, Mair, & George, 2004). Hardy, Beeton, and Pearson (2002) recommend that before tourism development begins, tourism planners should listen to the views of all stakeholders.

The tourism industry, government and local community have different goals and interests regarding STD (Timur & Getz, 2009). According to Timur and Getz (2009, p. 222), the local community and tourism operators share the common goal of economic and socio-cultural sustainability, while the local community, government officials and tourism industry share the common objective of sustainable resource use and protection aims. Moisey and McCool (2001) suggest sustainability can only ever be achieved in tourism developments once all stakeholder groups share common goals, and this requires the involvement of relevant local community members, government officials and tourism operators.

Most of the current theory associated with stakeholders and their role in tourism development is constructed from literature in the business management and public administration fields (Byrd, Bosley, & Dronberger, 2009). Much of the early research into stakeholders and stakeholder involvement focused on the organisation and the power of the stakeholder in public administration (Ansari & Phillips, 2001; Carmin, Darnall, & Mil-Homens, 2003; Steelman, 2001). Investigating different stakeholder groups and the importance of their perceptions of STD has also been a prominent theme in tourism research over the last decade (Andereck & Vogt, 2000; Davis & Morais, 2004; Haukeland, 2011; Holden, 2010; Ryan & Aicken, 2005).

Traditionally, tourism stakeholders were spilt into four main groups: visitors, the host community, tourism operators, and government officials (Goeldner & Ritchie, 2003). Early tourism-development studies (Allen, Hafer, Long, & Perdue, 1993; Lankford, 1994; Murphy, 1983; Pizam, Neumann, & Reichel, 1978) looked at only one of the four stakeholder groups; therefore, it is rare to find multiple stakeholder studies. Byrd et al. (2009, p. 694) note that in comparison to the large number of studies investigating individual stakeholder groups, there has been very little research undertaken to contrast the perceptions of different stakeholder groups. The incorporation of relevant stakeholder perspectives is essential to guaranteeing effective tourism management plans (Battaglia, Daddi, & Rizzi, 2012, p. 196). Focusing research on individual stakeholder perspectives often obscures the complexities existing between groups, whereas investigating multiple groups allows a comprehensive view of the big picture.

Stewart and Draper (2007, p. 7) note that there are an increasing number of methods available for investigating the perceptions of local stakeholders within a STD approach. In general, investigations into stakeholder perceptions have followed deductive frameworks, focusing on how individuals respond to predetermined questions and how attitudes relate to external and internal variables (Fredline & Faulkner, 2002). Previous work by Hardy (2005), Hasse and Milne (2005) and Stewart, Jacobson, and Draper (2008) has shown that combining interpretive research approaches helps in understanding the perceptions of local stakeholders affected by tourism development.

Stakeholder collaboration in sustainable tourism development

It is well documented that partnership building and collaboration are vital elements in enabling the sustainable development of tourism destinations (Moswete, Thapa, & Child, 2012; Nault & Stapleton, 2011; Timur & Getz, 2008). Pretty (1995) asserts that collaboration leads to the pooling of knowledge, expertise, capital and other resources, greater coordination of appropriate policies, increased approval of policies, and more successful implementation. The stakeholders remain autonomous with independent decision-making powers regardless of the group, within a framework of rules. Collaborative communication is generally a relatively formal process involving regular, face-to-face exchange of ideas, distinguishing it from other forms of participation (Carr, Selin, & Schuett, 1998). Maitland (2002) concludes that the creation and development of trust is an important goal for successful partnerships.

Tourism collaborations differ depending on several factors: geographic extent; whether the collaboration is voluntary or legislated; the balance of power between stakeholders; and organisational diversity and size all affect the outcomes of sustainable development (Selin, 2000). Kannapa (2011, p. 72) explains that tourism collaboration in developing nations requires greater levels of involvement from stakeholders than in developed countries, through increased dialogue, exchange of ideas and views and sharing lessons learnt, enabling them to better understand and effectively participate in local tourism development.

Selin and Chavez (1995, p. 848) conceived an 'evolutionary model of tourism partnerships,' which is helpful to determine how developed tourism collaboration is within a destination or community. The model consists of five stages: antecedents (initial circumstances); problem setting; direction setting; structuring; and outcomes (i.e. impacts on the domain). They suggest that tourism partnerships begin in a context of environmental forces and evolve sequentially through problem setting, direction setting and structuring phases.

There are some inherent difficulties with partnership approaches to tourism development. For example, the involvement of diverse stakeholders often requires regular meetings (Bramwell & Lane, 2000) and has the potential to be a complex prolonged process (Wray, 2011). Byrd (2007) adds that planners often do not properly collect opinions from all stakeholders involved in the decision-making process, reducing the success of the planning process. Hardy and Phillips (1998) add that the blocking of collaboration by an unequal power relationship between stakeholders can lead to conflicts between stakeholders resulting in unsuccessful tourism development.

Hall and Jenkins (1995) note that in many circumstances stakeholders may refuse to work with others as it could diminish their own influence, or they may distrust other groups. De Araujo and Bramwell (2002, p. 1139) add that in many tourism settings in less developed countries, there may be no practice of multiple stakeholder participation in decision-making. Previous studies have shown that the participation approach developed for First World countries may be unsuccessful in the socio-economic, cultural, administrative or political situation of a less developed country (Marzuki, 2008; Nault & Stapleton, 2011; Tosun, 2006).

There are greater challenges for those in economically less developed countries who wish to practise a collaborative tourism-development process compared with those in the developed world (Ashley, Roe, & Goodwin, 2001; Desai, 1996; Few, 2000). Recent studies by Lamberti, Noci, Guo, and Zhu (2011) in Shanghai, China, and Aref (2011) in Shiraz, Iran, found that restricted access to decision-making and a lack of resources, knowledge, skills and education were all barriers to community participation in economically less developed countries.

An administrative challenge occurring in many developing countries is that centralised-government controls can mean there is minimal local government influence, resulting in no incorporation of local community perspectives into management schemes (Tosun, 2006; Ying & Zhou, 2007). In addition, intricate bureaucracies and related jealousies within government splinter tourism development and hinder coordinated policy-making. Businessmen often get special treatment from powerful patrons, and poor communities generally do not have the time to be involved in tourism planning as they are busy working in order to survive (De Araujo & Bramwell, 2002, p. 1141).

De Araujo and Bramwell (2002, p. 1140) believe that there are a number of aspects of tourism-development partnerships that require further investigation. The focus needs to be on collaboration at a regional scale, as most previous studies have examined community or local-level collaboration (Jamal & Getz, 1999; Reed, 2000; as cited in De Araujo & Bramwell, 2002). Milne and Ateljevic (2001) add that understanding the global–local nexus is paramount to building constructive tourism partnerships and a need for embracement by tourism researchers.

The collaborative advantage

> The world of collaboration. It is a world in which it is possible to feel inspired. Almost anything is, in principle, possible through collaboration since you are not limited by your own resources and expertise. (Huxham & Vangen, 2005, p. 3)

Huxham and Vangen (2005, p. 3), in their breakdown of collaborative arrangements (not in the context of tourism), use the term 'collaborative advantage' to describe when something unusually creative is formed (for example, a goal is achieved), which no one group could have produced by itself. They add that through the collaboration each group is able to accomplish its own objectives better than it could alone. The purposes of collaboration are in general to advance a strategic and shared vision, with the overall objectives of increasing access to resources, sharing of risk, improving efficiency and seamlessness, and assisting learning (Huxham & Vangen, 2005, pp. 5–7). Hammer (2010, p. 209) adds that collaborative processes 'get the whole system in the room,' and are well suited to the complex conditions associated with community visioning and planning settings.

Huxham and Vangen (2005) explain their theory of collaborative advantage as a theme-based approach in which stakeholders identify issues that strain collaboration. The themes raised by stakeholders at this stage are generally specific to one group. Once initial themes are identified, the complexities of the collaboration become clearer. The final stage is a deeper examination of the data, revealing underlying themes that are common to many groups – in other words, 'cross-cutting' themes, mentioned in the discussion chapter of this study. In this last phase, tensions that pose barriers to collaboration are uncovered, and the nature of these tensions emerge. At the time of writing this study, the theory of collaborative advantage has rarely been applied to tourism research, and there are currently limited examples of it being utilised in a less developed nation context.

Despite collaboration being a recognised problem in tourism policy, research attempts have been hampered by a lack of cohesive thinking and the complexities involved in collaboration in this area. Devine, Boyle, and Boyd (2011) applied the theory of collaborative advantage to sports tourism in Northern Ireland. Devine et al. (2011, p. 37) found the approach helpful to explain inter-organisational relationships; however, they noted that no two collaborative settings are the same, that the themes overlap and are interlinked, and that collaboration is both complex and unpredictable.

Case study area: Mentawai Islands, Indonesia

The Mentawai Islands are situated between 85 and 135 kilometres off the coast of West Sumatra, Indonesia, positioned between 98°55′ East, 1°20′ South and 100°20′ East, 3°00′ South (Figure 1). The archipelago has four islands – Siberut, Sipora, North and South Pagai – which have a combined land area of about 7000 square kilometres (Whittaker, 2006). Siberut, the northernmost and largest of the Mentawai Islands, has a huge rainforest in the interior and a maze of inland waterways, which support diverse wildlife. Sipora, the smallest island, is between Siberut and North Pagai islands. It is home to the administrative capital, Tuapejat, and is the base of most government activity and industry for the Mentawais. North and South Pagai, the southernmost islands, are divided by a narrow strait and are the most remote of all the island group, being linked to the other islands via a weekly ferry service.

The Mentawai Islands have an estimated population of 70,000 people, comprising of up of 90% indigenous Mentawai (SurfAid International, 2008). There are, however, many other ethnic groups present in the Mentawai Islands, most notably Minangkabau, Javanese and Batak (Persoon, 2003). The major towns on the four main islands service the outlying villages. On Sundays, the local food markets draw a flotilla of small dugouts and motor boats from all over the island, with villagers intent on buying their fresh fruit and vegetables imported from mainland Sumatra. The larger towns are very similar to elsewhere in Indonesia. They have paved roads with hundreds of scooters and small buses rushing

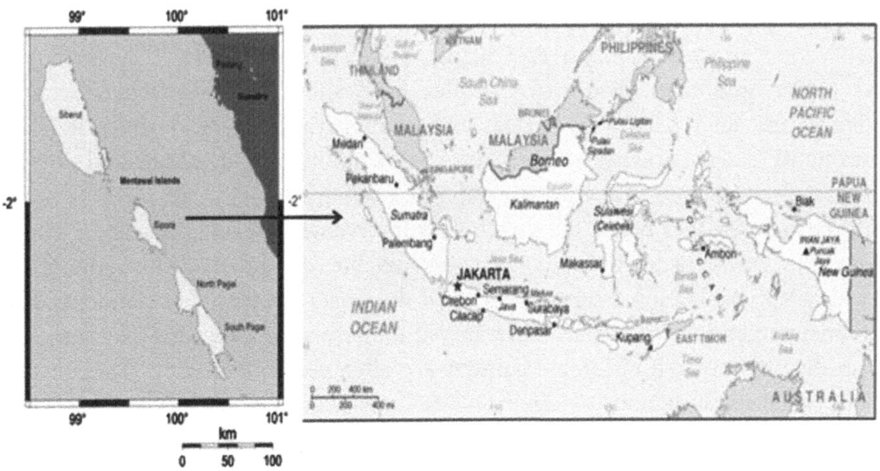

Figure 1. Map of the Mentawai Islands, Indonesia.

around, Internet cafes with children playing games, large mosques and diverse food eateries full of spicy Padang food.

The majority of inland villages are a stark contrast from their coastal counterparts, with the Mentawai people living in traditional small settlements (uma) along the banks of rivers within the dense jungle. Mentawai people in these isolated areas live a largely hunter-and-gatherer lifestyle; however, they have domesticated chickens and pigs and cultivated perennial crops, with sago, banana and taro being the staple foods. Employment opportunities in these traditional villages are scarce. The economy is agriculture based, with bananas and copra being the main exports, which are ferried to Padang on mainland Sumatra.

The Mentawai Islands are home to some of the finest waves for surfing to be found anywhere in the world, created by a perfect combination of long-range swell, light offshore winds and coral reefs. The waves break in very shallow water, creating long, hollow, high-performance waves, allowing surfers to complete multiple and varied manoeuvres. What makes the Mentawais even more marketable is that there are such a large number of good waves in close vicinity, reducing crowding. The majority of surfers who visit the Mentawai Islands stay at surf resorts or on board charter boats. The surf resorts are located in isolated natural environments, areas that have extremely sensitive ecosystems and less-developed indigenous communities, and are vulnerable to negative socio-cultural and environmental impacts from surf tourism (Buckley, 2002).

Methods

In total, the researcher spent three-and-a-half months conducting interviews with relevant stakeholders of the Mentawai Surf tourism Industry. Between 8 July and the 25 September 2010, 108 in depth community interviews were completed. Understanding and respecting the local culture was imperative in reducing barriers between the researcher and the study subjects. They were initially suspicious of an outsider conducting research in their local area. Having some awareness of cultural customs and traditions builds trust and respect between stakeholders and the researcher (Tchacos & Vallance, 2004). Engaging with the relevant stakeholders was essential, as their cooperation and input was vital to the success of the project. As an unknown researcher in the Mentawai Islands, his 'getting-in' or 'gaining the acceptance of the people being studied' was a vital first move in the data-collection process (Lofland & Lofland, 1984, p. 20). Clark, Riley, Wilkie, and Wood (1998) add that historical examples have shown that to achieve the most effective results, the researcher often has to gain approval from various individuals such as community leaders, understand local conditions, traditions and culture, and assess situations in order to identify the most appropriate respondents.

Before the researcher began the stakeholder interviews, several 'gatekeepers' (Burgess, 1991, p. 47) were contacted in order to identify suitable respondents. After gaining access, it was necessary to develop a rapport with and the trust of potential respondents. Fortuitously, many operators and locals remembered the researcher from his pilot visit. Outsiders conducting research in foreign cultures are often confronted with trust concerns as they can be viewed suspiciously (Jobbin, 2004). Cole (2004) discusses at length the importance of gaining the trust and building rapport with research respondents in isolated Indonesian communities, and that improved trust leads to richer insights and deeper understandings.

Local stakeholders were identified through these gatekeepers, observation and informal discussions. The process generally entailed casual walks through villages and discussions with prominent community members, operators and government officials about who might be appropriate stakeholders. When the most suitable people were identified, the researcher approached the prospective participants to introduce himself and the study. The researcher then gave them a participant information sheet which outlined the objectives of the study. After reading the sheet, they were required to sign a consent form prior to the interview commencing. Interviews were conducted on a one-to-one basis with the participant's permission.

Interviews were semi-structured, and guided by an interview schedule. The use of open-ended questions in the interviews allowed the views of the participants to emerge naturally, and not be influenced by the researcher (Glaser & Strauss, 2012). The researcher also frequently had to ask interviewees to explain themselves in more detail, check words and meanings with his translators, and trust his intuition regarding what questions were too sensitive or inappropriate. The interview questions investigated the participants' perspectives on community participation in the surf tourism industry and methods of improving local participation.

The length of the interviews varied, depending on the availability of the participant, the location and the information that needed to be collected; but interviews typically lasted approximately 60 minutes. In the field, the researcher found he had little control over the style of the interview, with the direction more dictated by the level of rapport between the participant and himself. The informal interviewing style, the semi-structured questions and relaxed atmosphere were all designed to make participants comfortable and to facilitate genuine responses to the questions they were being asked (Rubin & Rubin, 2005; Wengraf, 2001).

The text from the interviews was transcribed then analysed using NVivo (NVivo, 2010). Codes were assigned to meaningful statements that were central to these community participation dimensions, the goal being to generate the widest collection of codes possible. The large number of codes were then grouped to form conceptual value clusters, allowing similar instances of phenomena to be compared and contrasted and, where similar, to check they had been coded consistently (Douglas, 2003, p. 47). The process of constant comparison allowed for increased confidence and assurance in the previously generated codes (Glaser & Strauss, 2012). Various sub dimensions of community participation that were identified during the coding process were further explored.

Results

Industry stakeholders

Collaboration between the stakeholders (inter-group networking) was low, with just 17% of participants having regular collaboration with other groups. Involvement in the tourism planning process by operators, government officials and other stakeholders was similarly low, with only 19% of participants being involved in the regional tourism planning process. Surf charter boat operators had the lowest involvement in the tourism planning process, with many highlighting they were in the Mentawais to surf and take tourists on surf trips and had no desire to be involved in the bureaucracy of the industry. One comment made by a surf charter operator with a long history in the area emphasised that there was no forum or regularly scheduled meetings to discuss concerns, resulting in stressed relationships:

> "The greatest challenge for stakeholders is to be able to sit down, work together and agree as an industry, be supportive of each other. It seems too fractious." Australian, male, aged 50–59 years, charter boat operator

The general response from participants was that there is limited regular collaboration between different stakeholder groups. Boat operators pointed out, however, that they often talked over concerns amongst themselves in an informal setting such as out on the ocean or over a beer at a bar. Collaboration between community and operators seemed to be irregular and random, while collaboration between operators and government officials was avoided unless necessary.

> "Only really have regular collaboration with the other boat operators. Generally only communicate with community or government when there is an issue." Australian, male, aged 40–49 years, charter boat operator

> "The only regular collaboration we have as operators is with the owners and captains of other surf charter boats." Japanese, female, aged 30–39 years, charter boat operator

Three surf charter boat operators said their association – the West Sumatran Charter Boat Association (AKSSB) – looked after collaboration with other stakeholders. Similarly, two surf resort operators believed that their local Indonesian business partner (a requirement under Indonesian law) was effective in articulating tourism interests on their behalf so they felt they did not need to be directly involved.

> "I don't have direct collaboration with other stakeholders, but the West Sumatran Charter Boat Association tries to communicate my thoughts with other stakeholders." Australian, male, aged 40–49 years, charter boat operator

Another surf charter boat operator considered that the current tourism association (MMTA) did not fairly facilitate discussion on important issues such as taxes, anchorages and management strategies, and that there needed to be a holistic collaboration process where all voices were heard. In reality, creating a platform where all stakeholders are able to articulate their views would be almost impossible in the Mentawais due to the geographic isolation of certain stakeholders and the logistics of getting all groups in one place at one time.

> "There is not a committee or forum to discuss issues; there needs to be a tourism association with an even playing field." Australian, male, aged 40–49 years, resort operator

Comments from Mentawai government officials indicated that their involvement in the tourism planning process was low, but that they wished to be involved and learn about the process. Tourism officers expressed a need to have more dialogue with the host communities, but recognised it was logistically difficult because of the remote location of many villages.

> "I am not involved in the tourism planning process but I would like to know what goes on and how it all works." Indonesian, male, aged 30–39 years, government official

> "I have only limited collaboration with local operators not the whole industry, but I would like more collaboration with the local community but it is difficult. I think that the industry will benefit from it." Indonesian, male, aged 40–49 years, government official

One interviewee stated he was extremely enthusiastic about getting involved in tourism planning and had been trying unsuccessfully for five years. In 2006 the government asked industry stakeholders to submit ideas for an upcoming forum. However, after the meeting, there had been no contact or feedback regarding the outcome, and many of their suggestions had been overlooked.

> "There is no involvement in tourism planning; I have tried for over half a decade." American, male, aged 30–39 years, resort operator

All participants interviewed believed their village had not been involved in the tourism planning process. During his fieldwork, the researcher was frequently told about their community's attempts to become involved in the process, and that they wished to have an avenue to express their opinions on the future direction of that development. One interviewee from Mapadegat gave information to the Tourism Department outlining his thoughts on tourism in the region but had heard nothing back until over three years later. Three community members interviewed also expressed their eagerness to be involved in the tourism planning process. In isolated indigenous communities such as the Mentawai Islands, tourism planning is even more crucial because the strong traditions and cultural values are sensitive to Western influence.

> "I want the government to work with me; I want to get involved and learn the plan." male, aged 30–39 years, homestay operator, Mapadegat

> "No, we are not involved with tourism planning but myself and many other people from Mapadegat wish to be. I have given them a tourism proposal for our village and the surrounding area but have not heard back from them for a very long time." male, aged 40–49 years, village administrator, Mapadegat

All those interviewed said there had been very limited tourism-related collaboration between locals and either operators or government officials. Collaboration between local communities and government stakeholders was sporadic and fragmented. Participants claimed if the government wanted something from them, they would contact the village administrator and organise a one-off community meeting, but they would not receive any feedback. This indicated that limited or no collaboration was occurring between the two stakeholder groups. Likewise in the villages with more tourists: villagers from Silabu and Katiet confirmed tourism collaboration with government was extremely limited.

> "No, there has been no contact with the government or operators concerning surf tourism. I can only remember just one meeting ever with government that was about tourism." male, aged 30–39 years, village officer, Silabu

> "No, we never have meetings with the government for tourism. Collaboration between our village and the government is very irregular; there have only been a few meetings that I can recall." male, aged 30–39 years, homestay operator, Katiet

Interviewees emphasised that collaboration with operators was poor, which led to ill feeling and stressed relations between the parties. Villagers from Ebay claimed operators of the neighbouring resort had imposed unfair rules on them, restricting them from meeting with the tourists staying at the surf camp and preventing them from using their traditional land. This assertion came from residents of villages with a foreign-owned surf resort within 500 metres of where other local homestay accommodation is located.

> "No, there has been no contact with operators. We have had talks with the local resort but not stable, many problems." male, aged 30–39 years, NGO employee, Katiet

> "The relationship between locals at our village and the resort looks good but actually not so good: we cannot build on our land; locals are not allowed on resort grounds and resort is angry if locals talk with guests. It's better if tourists stay in homestay in village as they can meet locals." male, aged 40–49 years, homestay operator, Ebay

Some shopkeepers said that although collaboration with operator and government stakeholder groups was limited or in many cases non-existent, there were still a number of informal community meetings and discussions occurring. This is an interesting point because our Western concept of collaboration involving large meetings with many different individuals is very different to how the local Mentawai appear to perceive it. Their view on collaboration seemed to be more intimate, involving lengthy discussions with a smaller group restricted to key local villagers such as the village administrator and prominent business owners. Community members said it was uncommon to have meetings with other stakeholder groups outside their immediate village.

> "No, there has been no government input of any kind but we do have meetings within our own village." male, aged 30–39 years, shopkeeper, Mapadegat

> "I talk with the Kepala Desa very often about tourism matters but he is the only person consult I with. I do not talk to other stakeholders except for tourists who come and buy goods from my shop." male, aged 30–39 years, shopkeeper, Katiet

Discussion

The findings indicate that stakeholders involved in the Mentawai surf tourism industry all believe there is a need for shared collaboration to aid current and future development of tourism. The current lack of an organised collaboration initiative, with only sporadic and uncoordinated collaboration between stakeholder groups, is detrimental to all participants in the industry. All stakeholders agreed that collaboration between groups was limited and could be improved. Weak collaboration between stakeholders reflects and reinforces low participation in tourism planning and management processes, especially for those from the local community.

Government officials were the least involved in all the inter-stakeholder collaborations, formally and informally. Direct collaboration between government representatives and operators was very low, with operators indicating that they avoided collaboration with government officials unless it was necessary. Boat operators felt their organisation (AKSSB) was a more effective means of articulating their concerns to the relevant government officials, so their membership of this umbrella organisation obviated any need for individuals to communicate directly with government officials. The researcher was able to attend an AKSSB meeting, when in transit in Padang between field trips. Members of the AKSSB told him that belonging to an association and acting as a collective group gave them more chance for their opinions to be heard by government. Surf charter boat operators are well represented in the membership of AKSSB; but few were at the meeting the researcher attended.

Charter boat and resort operators perceived their collaboration with local community differently: land-based resorts reported they had high levels of strong productive collaboration with the host population, whereas the majority of charter boat operators viewed their collaboration with local community as limited. However, two charter boat operators did claim they had developed good relationships with some locals over the years, and that they would make an effort to meet up with them each time they were in the area. In one case when the researcher was on board a charter boat conducting an interview, an old Mentawai fisherman from Silabu paddled out on his dugout and sold mud crabs; the captain said the fisherman had been doing it for years and that the guests loved the fresh seafood.

On the other hand, resort operators believed that their collaboration with the host population was relatively strong and productive. They said that they consult with community on a variety of issues such as land use, employment of local people, and cultural education in order to maintain strong working relationships between the two parties. Interestingly, the issue of fractured and stressed relations apparent in the community interviews failed to surface during interviews with the resort operators. Their responses perhaps masked the underlying tension between the two groups and the issue of restricted access to ancestral land.

The majority of community members interviewed had quite a different viewpoint. They said they had low levels of collaboration with land-based operations. For example, the management from the now-defunct HT Villas at Katiet village had had a number of clashes with locals regarding access to the beach and trespassing. Homestay operators highlighted that relations between locals and a nearby resort were not as good as they would appear to be on the surface. Villagers suggested that the resort operators would have their guests and other surf tourists believe everything was 'great' when in fact communication between them had ceased. This was due to the resort's policy of barring interaction between locals and guests and preventing locals' access to traditional resources on resort grounds.

One key finding from this section on resort operators and local community collaboration is that, although the two different groups had opposing viewpoints about their current levels of collaboration, both groups see a need for collaboration to be improved. Resort owners want to build on what they say are already high levels of collaboration with the local community, whereas local community say collaboration needs to be improved because their current collaboration with resort operators is strained and infrequent.

Government stakeholders recognised that their collaboration with the local community was low, but they were enthusiastic about increasing those levels of collaboration. Inadequate staffing and financing within local government are ever-present problems, with one high-level member of the Mentawai Tourism Department admitting there was a lack of experienced and well-trained individuals who would be able to facilitate effective collaboration between the various stakeholder groups. The same official also noted there were insufficient financial resources to implement successful tourism collaboration. Moreover, collaboration is weak even within the local government because the inter-departmental collaboration systems between the three departments involved in surf tourism in the Mentawai Islands (tourism, marine and transport) are poor.

Interviewees in all study communities indicated that their collaboration with government stakeholders in the industry was sporadic, informal and largely ineffective. Collaboration between community and the local Mentawai government has been dogged by corruption. Locals are distrustful, even fearful of those in positions of authority. A long-held tradition of suspicion and mistrust has further separated the host population from government officials, restricting the success of collaboration attempts. In general, government collaboration with community had been limited to one-off meetings, following which local participants rarely received any feedback.

It is clear that collaboration between government and community in the Mentawai Islands is almost non-existent. Collaboration between the two stakeholder groups needs to go beyond sporadic uncoordinated meetings and become part of a flexible and dynamic process that evolves over time, enabling community stakeholders to address their concerns. Nault and Stapleton (2011, p. 695) note that close and sustained collaboration between trusted community leaders and knowledgeable outside stakeholders is required for the long-term viability of any community-based tourism development in a remote area.

A significant finding of this study is that operators, community members and government officials want increased levels of collaboration with each other. Local community members desire greater involvement in the tourism planning and development collaboration process, while government officials showed enthusiasm for increasing their levels of collaboration. The interviews also revealed that there must be some reconciliation between the parties before progress can be made in improving collaboration between the two groups. Specifically, government officials need to address the historical distrust felt by local communities.

Tourism forums (such as an annual summit in Tuapejat, the capital of the Mentawai Islands) would provide a platform for effective collaboration, where relevant tourism stakeholders could discuss concerns about surf tourism development, management and marketing. A successful tourism forum in the Mentawai Islands would require identification and representation of all appropriate stakeholders involved in the surf tourism industry. Studies from other developing countries have shown that collaboration between the local community and other tourism industry stakeholders can relieve tension (Jamal & Getz, 1999) and lead to rich insights that can guide future development (Grybovych & Hafermann, 2010, p. 354; Spencer, 2010, p. 684).

Forums that incorporate the multiple stakeholder approach allow the views of all relevant community, government, operators and NGO groups in the surf tourism industry to be fully communicated (De Araujo & Bramwell, 2002; Larson & Poudyal, 2012). Greater recognition of local perspectives would help other stakeholders understand the challenges they face in becoming more actively engaged in the industry (Tosun & Timothy, 2003). Tourism forums could potentially assist previously marginalised communities to achieve a sense of empowerment (Okazaki, 2008) and help improve communication between themselves, local government and surf operators.

An example of a successful tourism forum is one which was held in rural Eastern Indonesia in 2003 when Cole organised a one-day tourism seminar with local tourism industry stakeholders as part of the dissemination of her study findings (Cole, 2006). Local stakeholders involved in the tourism industry were present, including 15 villagers, the local guides, four officials from the Department of Tourism and three members from the Department of Education and Culture. The high stakeholder turnout was achieved through official invitations from the Department of Tourism, and villagers' travel expenditures were covered by the government. The outcome was very positive, with the Department of Tourism agreeing it was a useful process and should be repeated. The public forum gave the villagers a feeling of empowerment, allowing them to speak out against the government and to raise ideas and issues in a public forum, an important first step in the transfer of knowledge to the local community (Cole, 2006, pp. 638–640).

Although in theory this type of tourism forum sounds achievable, in a geographically isolated region such as the Mentawai Islands, the logistics and expenses involved in getting all representative stakeholders together are significant obstacles. Other impediments include a lack of experienced people available to facilitate a successful forum and the existing mistrust between community and government, which has led to difficulties in reaching mutually agreed-upon outcomes. To make a large multi-stakeholder tourism forum feasible would require both the Sumatran and Mentawai governments to take a leading role in supporting stakeholder collaboration, and provide both financial and managerial assistance.

Conclusion

Distinct contributions have been made to the growing literature on the value of exploring multiple stakeholder collaboration in sustainable tourism development (Aref, 2011; Kannapa, 2011; Nault & Stapleton, 2011). Adopting a multiple stakeholder perspective and examining the views of both major stakeholder groups (government, operators, community and tourists) and non-traditional tourism stakeholder groups (NGO employees, surf journalists) sheds new light on our current understanding of the importance of collaboration in sustainable tourism development. The ability to have the power to combine all the detailed data of the numerous stakeholder groups meant that the smaller valuable pieces of information assisted in forming a comprehensive and accurate picture of surfing tourism development in the Mentawai Islands.

This study echoes previous research (Andereck, Valentine, Knopf, & Vogt, 2005; Stewart & Draper, 2007; Tosun, 2006; Yang, Ryan, & Zhang, 2013) that has emphasised the importance of adequately incorporating the views of the local community in the research approach. This was crucial in the Mentawai context, as the Islands' remote location and the locals' limited tourism experience have meant that they have been largely excluded from the decision-making processes of the surf tourism industry. The study highlighted the importance of understanding local stakeholder viewpoints in a complex cultural and political context, allowing important commonalities and differences between groups to surface during the analysis process.

Distinct contributions are also made to the growing literature on the value of exploring multiple stakeholder collaboration in sustainable tourism development (Aref, 2011; Kannapa, 2011; Nault & Stapleton, 2011). This study showed that collaborative theories designed for non-tourism sectors (Devine et al., 2011; Huxham & Vangen, 2005) could be applied to tourism in the context of the Mentawai Islands. The major contribution to stakeholder collaboration theory was that it not only provided insights into how different stakeholders perceived development, but also took the analysis to a deeper level and found common ground upon which stakeholders could collaborate and exchange information and thereby develop a way forward.

This study has illustrated the complex challenges that stakeholders face in attempting to enhance the sustainability of surf tourism development in the Mentawai Islands. Despite the many obstacles facing various sectors of the industry, one central, common theme should be noted: all stakeholders showed enthusiasm to work together to ensure the Mentawai Islands became a more sustainable destination. One suggestion stemming from this contribution is that a way forward can be developed through stakeholders finding common ground after collaborating and exchanging information, for example by the implementation of an annual tourism forum at which relevant stakeholders could convey their viewpoints regarding surf tourism development on an equal platform. Staging such a forum in a neutral setting with third-party facilitation would help previously marginalised stakeholders to feel more comfortable about openly discussing their concerns regarding tourism development. Forums conducted in this fashion would act as a vehicle for opening communication channels, helping to reconcile historical distrust and foster improved relationships between groups. The forums would provide direction for a sustainable tourism strategy, highlighting the common ground between stakeholders.

It is recommended that future studies examine the feasibility of implementing ways for stakeholders to collaborate effectively with each other and the practicalities of developing productive 'tourism forums' that included all relevant stakeholders. These tourism forums would provide an appropriate platform for stakeholders to establish common ground and develop the most suitable way forward for their industry. It is vital that research now be focused on the mechanisms that could facilitate such forums. This paper has pointed to a range of constraints including geographic isolation, transportation, corruption and lack of trust that will be challenging to overcome, but it has also highlighted important points of agreement and common interest that can be explored and developed further. Questions that need to be answered include: How might these forums be operationalised? Who would drive them? How would they be funded?

Disclosure statement

No potential conflict of interest was reported by the author.

References

Allen, L. H., Hafer, P., Long, R., & Perdue, R. (1993). Rural residents' attitudes toward recreation and tourism development. *Journal of Travel Research*, *31*(4), 27–33.

Andereck, K. L., Valentine, K. M., Knopf, R. C., & Vogt, C. A. (2005). Residents' perceptions of community tourism impacts. *Annals of Tourism Research*, *32*(4), 1056–1076.

Andereck, K. L., & Vogt, C. A. (2000). The relationship between residents' attitudes toward tourism and tourism development options. *Journal of Travel Research*, *39*, 27–36.

Ansari, W. E., & Phillips, C. J. (2001). Interprofessional collaboration: A stakeholder approach to evaluation of voluntary participation in community partnerships. *Journal of Interprofessional Care*, *15*(4), 351–368.

Aref, F. (2011). Barriers to community capacity building for tourism development in communities in Shiraz, Iran. *Journal of Sustainable Tourism, 19*(3), 347–359.

Ashley, C., Roe, D., & Goodwin, H. (2001). *Pro-poor tourism strategies: Making tourism work for the poor. A review of experience.* London: Overseas Development Institute.

Battaglia, M., Daddi, T., & Rizzi, F. (2012). Sustainable tourism planning and consultation: Evidence from the project INTER.ECO.TUR. *European Planning Studies, 20*(2), 193–211.

Bramwell, B., & Lane, B. (2000). Collaboration and partnerships in tourism planning. In B. Bramwell, & B. Lane (Eds.), *Tourism collaboration and partnerships: Politics, practice and sustainability* (pp. 1–19). Clevedon: Channel View.

Buckley, R. (2002). Surf tourism and sustainable development in indo-Pacific Islands. II. Recreational capacity management and case study. *Journal of Sustainable Tourism, 10*(5), 425–442.

Burgess, R. G. (1991). Sponsors, gatekeepers, members, and friends: Access in educational settings. In W. B. Shaffir, & R. A. Stebbins (Eds.), *Experiencing fieldwork: An inside view of qualitative research* (pp. 43–52). Newbury Park, CA: Sage Publications.

Burns, P. (2003). Tourism planning: A third way? *Annals of Tourism Research, 31*, 24–43.

Byrd, E. T. (2007). Stakeholders in sustainable tourism development and their roles: Applying stakeholder theory to sustainable tourism development. *Tourism Review, 62*(2), 6–13.

Byrd, E. T., Bosley, H. E., & Dronberger, M. G. (2009). Comparisons of stakeholder perceptions of tourism impacts in rural eastern North Carolina. *Tourism Management, 30*, 693–703.

Carmin, J., Darnall, N., & Mil-Homens, J. (2003). Stakeholder involvement in the design of U.S. voluntary environmental programs: Does sponsorship matter? *Policy Studies Journal, 31*(4), 527–543.

Carr, D., Selin, S., & Schuett, M. (1998). Managing public forests: Understanding the role of collaborative planning. *Environmental Management, 22*, 767–776.

Clark, M., Riley, M., Wilkie, E., & Wood, R. (1998). *Researching and writing dissertations in hospitality and tourism.* London: International Thomson Business Press.

Cole, S. (2004). Longitudinal research in Eastern Indonesia. In J. Phillimore & L. Goodson (Eds.), *Qualitative research in tourism. Ontologies, epistemologies and methodologies* (pp. 292–310). London: Routledge.

Cole, S. (2006). Information and empowerment: The keys to achieving sustainable tourism. *Journal of Sustainable Tourism, 14*(6), 629–644.

Davis, J. S., & Morais, D. P. (2004). Factions and enclaves: Small towns and socially unsustainable tourism development. *Journal of Travel Research, 43*, 3–10.

De Araujo, L. M., & Bramwell, B. (2002). Partnership and regional tourism in Brazil. *Annals of Tourism Research, 29*(4), 1138–1164.

Desai, V. (1996). Access to power and participation. *Third World Planning Review, 18*, 217–242.

Devine, A., Boyle, E., & Boyd, S. (2011). Towards a theory of collaborative advantage for the sports tourism policy arena. *International Journal of Public Sector Management, 24*(1), 23–41.

Donaldson, T., & Preston, L. E. (1995). The stakeholder theory of the corporation: Concepts, evidence, and implications. *Academy of Management Review, 20*(1), 65–91.

Douglas, D. (2003). Grounded theories of management: A methodological review. *Management Research News, 26*(5), 44–52.

Few, R. (2000). Conservation, participation and power: Protected-area planning in the coastal zone of Belize. *Journal of Planning Education and Research, 19*, 401–408.

Fredline, E., & Faulkner, B. (2002). Residents' reactions to the staging of major motorsport events within their communities: A cluster analysis. *Event Management, 7*, 103–114.

Freeman, R. E. (1984). *Strategic management: A stakeholder approach.* Boston: Pitman.

Glaser, B., & Strauss, A. (2012). *The discovery of grounded theory: Strategies for qualitative research* (7th ed.). New York, NY: Aldine.

Goeldner, C. R., & Ritchie, J. R. B. (2003). *Tourism: Principles, practices, philosophies* (9th ed.). Hoboken, NJ: John Wiley.

Gough, V. (1999). *Assessing the economic effects of recreation facility development – Proposed artificial surfing reef, Mount Maunganui, New Zealand.* Waikato: University of Waikato.

Gray, B. (1985). Conditions facilitating interorganizational collaboration. *Human Relations, 38*(10), 911–936.

Gray, B. (1989). *Collaboration: Finding common ground for multi-party problems.* San Francisco: Jossey-Bass.

Grybovych, O., & Hafermann, D. (2010). Sustainable practices of community tourism planning: Lessons from a remote community. *Community Development, 41*(3), 354–369.

Hall, C., & Jenkins, J. (1995). *Tourism and public policy.* London: Sage.

Hammer, J. M. (2010). Large group interventions as a tool for community visioning and planning. *Community Development: Journal of the Community Development Society, 41*(2), 209–222.

Hardy, A. (2001). *A troubled paradise: Stakeholder perceptions of tourism in the Daintree Region of Far North Queensland, Australia.* PhD thesis, University of Queensland, Brisbane.

Hardy, A. (2005). Using grounded theory to explore stakeholder perceptions of tourism. *Journal of Tourism and Cultural Change, 3*(2), 108–133.

Hardy, A., Beeton, R., & Pearson, L. (2002). Sustainable tourism: An overview of the concept and its position in relation to conceptualisations of tourism. *Journal of Sustainable Tourism, 10*(6), 475–496.

Hardy, C., & Phillips, N. (1998). Strategies of engagement: Lessons from the critical examination of collaboration and conflict in an interorganizational domain. *Organization Science, 9*(2), 217–230.

Hasse, J., & Milne, S. (2005). Participatory approaches and geographical information systems (PAGIS) in tourism planning. *Tourism Geographies, 7,* 272–289.

Haukeland, J. V. (2011). Tourism stakeholders' perceptions of national park management in Norway. *Journal of Sustainable Tourism, 19*(2), 133–153.

Holden, A. (2010). Exploring stakeholders' perceptions of sustainable tourism development in the Annapurna conservation area: Issues and challenge. *Tourism and Hospitality Planning & Development, 7*(4), 337–351.

Huxham, C., & Vangen, S. (2005). *Managing to collaborate. The theory and practice of collaborative advantage.* London: Routledge.

Jamal, T., & Getz, D. (1999). Community roundtables for tourism-related conflicts: The dialectics of consensus and process structures. *Journal of Sustainable Tourism, 7*(3–4), 290–313.

Jobbin, G. (2004). Translators: Cross-cultural issues. In J. Phillimore, & L. Goodson (Eds.), *Qualitative research in tourism. Ontologies, epistemologies and methodologies* (pp. 311–323). London: Routledge.

Kannapa, P. (2011). Participatory management process in local tourism development: A case study on fisherman village on Samui Island, Thailand. *Asia Pacific Journal of Tourism Research, 16*(1), 57–73.

Lamberti, L., Noci, G., Guo, J., & Zhu, S. (2011). Mega-events as drivers of community participation in developing countries: The case of Shanghai World Expo. *Tourism Management, 32*(6), 1474–1483.

Lankford, S. V. (1994). Attitudes and perceptions toward tourism and rural regional development. *Journal of Travel Research, 32*(3), 35–43.

Larson, L. R., & Poudyal, N. C. (2012). Developing sustainable tourism through adaptive resource management: A case study of Machu Picchu, Peru. *Journal of Sustainable Tourism, 20*(7), 917–938.

Lazarow, N. (2007). The value of coastal recreational resources: A case study approach to examine the value of recreational surfing to specific locales. *Journal of Coastal Research,* SI 50 Proceedings of the 9th International Coastal Symposium, Gold Coast, Australia, pp.12–20.

Lazarow, N., Miller, M., & Blackwell, B. (2007). *Dropping in a case study approach to understanding the socio-economic impact of recreational surfing and its value to the tourism industry.* Paper presented at the 5th International Coastal and Marine Congress, Auckland, New Zealand.

Lazarow, N., Miller, M., & Blackwell, B. (2008). The value of recreational surfing to society. *Tourism in Marine Environments, 5*(2–3), 145–158.

Lofland, J., & Lofland, L. H. (1984). *Analysing social settings: A guide to qualitative observation and analysis.* Belmont: Wadsworth Publishing Company.

Maitland, R. (2002). Creating successful partnerships in urban tourism destinations: The case of Cambridge. *Tourism, 3*(50), 293–302.
Martin, S. A., & Assenov, I. (2012). The genesis of a new body of sport tourism literature: A systematic review of surf tourism research (1997–2011). *Journal of Sport and Tourism, 17*(4), 257–287.
Marzuki, A. (2008). Decision making and community participation: A case study of the tourism industry in Langkawi. *Tourism, 56*(3), 227–241.
Milne, S., & Ateljevic, I. (2001). Tourism, economic development and the global-local nexus: Theory embracing complexity. *Tourism Geographies, 3*(4), 369–393.
Moisey, R. N., & McCool, S. F. (2001). Sustainable tourism in the 21st century: Lessons from the past: Challenges to address. In S. F. McCool & R. N. Moisey (Eds.), *Tourism, recreation and sustainability: Linking culture and the environment* (pp. 343-352). Wallingford, UK: CABI.
Moswete, N. N., Thapa, B., & Child, B. (2012). Attitudes and opinions of local and national public sector stakeholders towards Kgalagadi Transfrontier Park, Botswana. *International Journal of Sustainable Development and World Ecology, 19*(1), 67–80.
Murphy, P. E. (1983). Tourism as a community industry: An ecological model of tourism development. *Tourism Management, 4*, 180–193.
Nault, S., & Stapleton, P. (2011). The community participation process in ecotourism development: A case study of the community of Sogoog, Bayan-Ulgii, Mongolia. *Journal of Sustainable Tourism, 19*(6), 695–712.
Nelsen, C., Pendleton, L., & Vaughn, R. (2007). A socio-economic study of surfers at Trestles Beach. *Shore and Beach, 75*(4), 32–38.
Ntloko, N. J., & Swart, K. (2008). Sport tourism event impacts on the host community: A case study of Red Bull big wave Africa. *South African Journal for Research in Sport, Physical Education and Recreation, 30*(2), 79–93.
NVivo. (2010). NVivo qualitative data analysis software (Version 9). QSR International Pty Ltd.
Okazaki, E. (2008). A community-based tourism model: Its conception and use. *Journal of Sustainable Tourism, 16*(5), 511–529.
Persoon, G. A. (2003). Conflicts over trees and waves on Siberut Island. *Geografiska Annaler: Series B, Human Geography, 85*(4), 253–264.
Pizam, A., Neumann, Y., & Reichel, A. (1978). Dimensions of tourist satisfaction with a destination area. *Annals of Tourism Research, 5*(3), 314–322.
Ponting, J. (2007). The endless summer? The past, present and future of surfing tourism management in the Pacific. In I. McDonnell, S. Grabowski, & R. March (Eds.) *Tourism, past achievements, future challenges. Proceedings of the 17th Annual CAUTHE Conference* (pp. 1–9). Sydney, Australia: University of Technology.
Ponting, J. (2008). *Consuming Nirvana: An exploration of surfing tourist space*. Doctoral study, University of Technology, Sydney, Australia.
Ponting, J. (2009). Projecting paradise: The surf media and the hermeneutic circle in surfing tourism. *Tourism Analysis, 14*, 175–185.
Ponting, J., McDonald, M., & Wearing, S. (2005). De-constructing Wonderland: Surfing tourism in the Mentawai Islands, Indonesia. *Society and Leisure, 28*(1), 141–162.
Ponting, J., & O'Brien, D. (2013). Liberalizing Nirvana: An analysis of the consequences of common pool resource deregulation for the sustainability of Fiji's surf tourism industry. *Journal of Sustainable Tourism*, Article in press.
Pretty, J. (1995). The many interpretations of participation. *In Focus, 16*, 4–5.
Reed, M. (2000). Collaborative tourism planning as adaptive experiments in emergent tourism settings. In B. Bramwell, & B. Lane (Eds.), *Tourism collaboration and partnerships: Politics, practice and sustainability* (pp. 247–271). Clevedon: Channel View.
Reid, D. G., Mair, H., & George, W. (2004). Community tourism planning: A self-assessment instrument. *Annals of Tourism Research, 31*(3), 623–639.
Rubin, H. J., & Rubin, I. S. (2005). *Qualitative interviewing: The art of hearing data* (2nd ed.). Thousand Oaks, CA: Sage.

Ryan, C., & Aicken, M. (2005). *Indigenous tourism: The commodification and management of culture*. Amsterdam: Elsevier.

Selin, S. (2000). Developing a typology of sustainable tourism partnerships. In B. Bramwell, & B. Lane (Eds.), *Tourism collaboration and partnerships: Politics, practice and sustainability* (pp. 129–142). Clevedon: Channel View.

Selin, S., & Chavez, D. (1995). Developing an evolutionary tourism partnership model. *Annals of Tourism Research*, *22*, 844–856.

Spencer, D. (2010). Facilitating public participation in tourism planning on American Indian reservations: A case study involving the Nominal Group Technique. *Tourism Management*, *31*, 684–690.

Steelman, T. A. (2001). Elite and participatory policymaking: Finding balance in a case of national forest planning. *Policy Studies Journal*, *29*(1), 71–89.

Stewart, E. J., & Draper, D. (2007). A collaborative approach to understanding local stakeholder perceptions of tourism in Churchill, Manitoba (Canada). *Polar Geography*, *30*(1), 7–35.

Stewart, E. J., Jacobson, D., & Draper, D. (2008). Public participation geographic information system (PPGIS): Challenges of implementation in Churchill, Manitoba. *The Canadian Geographer / Le Géographe canadien*, *52*(3), 351–366.

SurfAid International. (2008). *Annual report 2008*. Retrieved August 29, 2010, from http://www.surfaidinternational.org/

Tchacos, E., & Vallance, R. J. (2004). *Research in aboriginal communities: Cultural sensitivity as a prerequisite*. Presented at Australian Association for Research in Education (AARE) November 28th–December 2nd 2004. The University of Melbourne.

Timur, S., & Getz, D. (2008). A network perspective on managing stakeholders for sustainable urban tourism. *International Journal of Contemporary Hospitality Management*, *20*(4), 445–461.

Timur, S., & Getz, D. (2009). Sustainable tourism development: How do destination stakeholders perceive sustainable urban tourism? *Sustainable Development*, *17*, 220–232.

Tosun, C. (2006). Expected nature of community participation in tourism development. *Tourism Management*, *27*, 493–504.

Tosun, C., & Timothy, D. (2003). Arguments for community participation in tourism development. *Journal of Tourism Studies*, *14*(2), 2–11.

Wengraf, T. (2001). *Qualitative research interviewing*. London: Sage.

Whittaker, D. J. (2006). A conservation action plan for the Mentawai Primates. *Primate Conservation*, *20*, 95–105.

Wray, M. (2011). Adopting and implementing a transactive approach to sustainable tourism planning: Translating theory into practice. *Journal of Sustainable Tourism*, *19*(4–5), 605–627.

Yang, J., Ryan, C., & Zhang, L. (2013). Social conflict in communities impacted by tourism. *Tourism Management*, *35*, 82–93.

Ying, T., & Zhou, Y. (2007). Community, government and external capitals in China's rural cultural tourism: A comparative study of two adjacent villages. *Tourism Management*, *28*(1), 96–107.

Envisioning Eden: the manufactured ecotourism environment of Singapore

Aaron Tham

ABSTRACT

The popularity of adopting ecotourism initiatives has led to a mushrooming of eco-precincts around the world. Arguably, ecotourism fulfils an ideological paradigm because of its inherent benefits to both natural and human environments. However, whilst theory has guided the justification for ecotourism and development of its principles, very little exists to steer how ecotourism should be unpacked across considerably varying dimensions of ecotourism across the globe. This research seeks to explain the notion of manufactured ecotourism from the perspective of Singapore. In a land-scarce country, ecotourism could hardly be conceivable in an environment where occupying physical space is expected to bring high economic returns on infrastructural investment. However, from a comparative case study approach, three specific exemplars of ecotourism developments have shown that the manufacturing of ecotourism has shifted the narrative of land use in Singapore and continues to transform the tourism landscape in the country. The outcomes of the research steer the directions of planning and stakeholder partnership for new destinations embarking on their ecotourism journey to 'Envision Eden'.

Introduction

The term ecotourism has attracted significant scholarly attention over the last two decades (Carrier & Macleod, 2005; Cater, 2006; Weaver & Lawton, 2007). This may be attributed to the growing concerns with tourism developments and the associated levels of stress it can place on the natural environment. Around the world, tourism has inadvertently led to some heightened impacts on the natural environment that has threatened the existence of some species of flora and fauna (Buckley, 2012; Gladstone, Curley, & Shokri, 2013; Tang, 2015).

Hence, ecotourism espouses an ideal environment that offers an alternative perspective to mainstream tourism development (Sharpley, 2006; Stronza, 2007). By shifting the trajectory from mass tourism to a model of sustainability, ecotourism resonates with the conservation of ecological precincts as well as the economic and socio-cultural well-being of local communities (Lai & Nepal, 2006; Tsaur, Lin, & Lin, 2006). Although the benefits of ecotourism are generally accepted, there has been a lack of consensus as to how to define ecotourism (Bjork, 2000; Blamey, 1997). Through an analysis of 85 iterations of the term ecotourism, Fennell (2001) found that the different definitions had shifted from a focus on the destination context towards overarching principles on conservation and sustainability. Likewise, Donohoe and Needham (2006) concluded that there has been a very broad interpretation of ecotourism definitions around the world. While having an inclusive approach is useful to guide managerial and operational practices around the world, the lack of a unifying definition can give rise to some inherent challenges when assessing the trajectories of ecotourism over time (Higham & Luck, 2002; Kruger, 2005). Such a limitation notwithstanding, this research subscribes to the view of Weaver (2005) in that ecotourism practices around the world occur across a hard to soft spectrum depending on the context that it represents. Yet, as Weaver (2005) has contended, such a perspective requires further clarification about how ecotourism should be interpreted based on its own context.

With this as a backdrop, this research aims to examine how ecotourism will feature in Singapore. Being one of the smallest nations in the world, the land-scarce environment privileges land use that generates significant returns on investment. However, within the last 12 months, the country has hastened its efforts for nature-based experiences, which has led to the groundbreaking initiatives for three case studies to be considered for ecotourism. This sudden amplification of ecotourism as a new tourism agenda lends a timely justification for this research to be undertaken as no precedence exists in Singapore.

Setting the scene

Literature on the tourism landscape in Singapore has focused primarily on a range of activities. These include food (Henderson, 2004; Hui, Wan, & Ho, 2007), shopping (Lew, 1992; Yeung, Wong, & Ko, 2004), heritage sites (Chang, 1999; Teo & Huang, 1995), medical tourism (Lee, 2010; Leng, 2010) and animal interactions, such as the Singapore Zoo (Tsai, 2012). Being at the hub of air networks in Asia, Singapore has traditionally been considered a transit point for flights between the East and West (Lohmann, Albers, Koch, & Pavlovich, 2009). However, the country has in the last decade invested heavily in tourism to attract greater visitor numbers. The focus on tourism is an outcome of the service economy as Singapore does not have significant raw materials that it can produce (Gylfason & Zoega, 2006). In this space, new tourist attractions and events have been created, including two casinos and a Formula One race (Henderson, Foo, Lim, & Yip, 2010; Yap, 2007). The overall aims of such initiatives are to harness the destination appeal of the country and ultimately reap benefits associated with extending visitors' average length of stay.

Incidentally, the catalyst for ecotourism in Singapore may have been exacerbated by other neighbouring countries that have targeted the 'green consumer'. For instance, Malaysia, Thailand and Indonesia have already had some successes with the development of ecotourism experiences (Chan & Baum, 2007; Galley & Clifton, 2004; Kontogeorgopoulos, 2004). While Singapore has been known as a garden city, few studies have been undertaken to determine its feasibility in developing ecotourism, especially when land is a valuable commodity. For instance, Wu (2012) contended that whilst ecotourism can have desirable outcomes for Singapore, uptake by local tourism operators has been relatively slow due to the lack of awareness and governmental support. Moreover, a longitudinal study by Litvin and Chiam (2014) also showed that the domestic demand for ecotourism did not improve over time and age segments.

Nonetheless, the turning point for ecotourism developments came about as part of the country's push for sustainability (Ooi, 2007). According to Newman (2014), the country is under pressure due to the rapid demands of a growing population and as such, findings ways to connect nature with the changing trends of society. For this reason, much attention has shifted to the nature belt. Three recent developments are the government's community engagement forum to redevelop Pulau Ubin (Salleh, 2014), the opening of Gardens by the Bay (Preece, 2012) and most notably, a successful push for the Singapore Botanic Gardens to obtain world heritage status (Siau, 2015).

In the context of ecotourism, some areas in Singapore have been given the approval to conduct relevant ecotourism experiences. This research will feature three such case studies – the Sister Island Marine Park (SIMP), Kampung Kampus and Kallang Riverside Eco-Park. These developments have only commenced within the last 12 months and are very much work in progress. Nonetheless, the willingness of local authorities to apportion land for ecotourism use is very much well-received and yet met with skepticism. This is because the site operators have neither significant financing nor possess any ecotourism expertise. Hence, it is envisaged that stakeholder involvement will be essential in order for desired ecotourism outcomes to come to fruition.

Stakeholder theory and ecotourism

Stakeholder theory is concerned with the management of business entities that resonate with the needs and values of any individual or group (Fletcher, 2009). Specifically, a stakeholder can have a direct impact on how the business operates, and also be affected by the directions and implications of the business (Jackson, 2007). In an ecotourism perspective, it is clearly evident that stakeholders encompass government, residents, ecotourism operators and other ancillary tourism and hospitality services such as accommodation and dining establishments.

Whilst there is an overall consensus about who the stakeholders are to any ecotourism development, there remains an imminent tension as to how ecotourism should be developed so that mutually agreeable goals and values are obtained (Garrod, 2003). In some studies, this tension has arisen from a pressing demand that ecotourism realises quick returns on investment and neglecting locally held customs and societal values (Brondo & Woods, 2007; Jamal, Borges, & Stronza, 2006; Liu & Bao, 2004). Other studies also allude to ecotourism developments expedited by foreign investors who may marginalise the needs of resident communities (Lawrence, Wickins, & Phillips, 1997; Medina, 2005). These circumstances have led Donohoe and Needham (2006) to conclude that whilst there is an agreement as to who the ecotourism stakeholders are, value appropriation of ecotourism outcomes remain a highly disputed issue.

The scope of stakeholder theory in ecotourism has been heavily skewed in the direction of developing tourism economies. For instance, Gurung and Scholz (2008) examined stakeholder interests in the context of Bhutan and highlighted the dangers of eroding societal

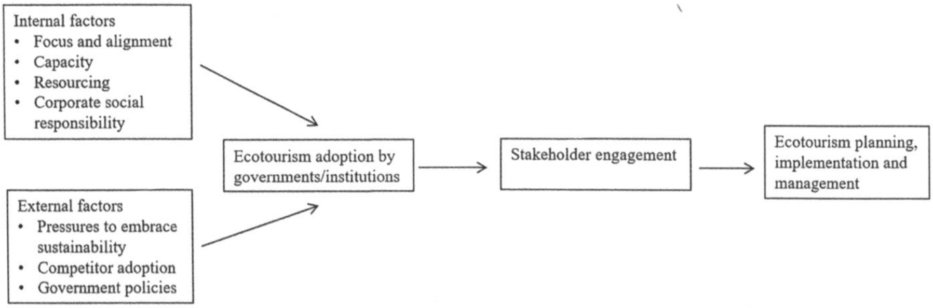

Figure 1. Conceptual model derived from Boyd and Butler (1996), Garrod (2003), Orams (1995); Wight (1993).

values. Such an outcome was also observed by Gui, Fang, and Liu (2004) in the context of China, because local communities had little knowledge of ecotourism in order to derive sustainable benefits. Yet, for many other developing countries, ecotourism is elevated to a status where it promises poverty alleviation and nature-based conservation (Caffyn & Jobbins, 2003; de Araujo & Bramwell, 1999; Gray & Campbell, 2007). These studies epitomise the stages within a revised model of ecotourism management proposed by Garrod (2003). In his model, Garrod (2003) highlights a consultative approach to ecotourism where stakeholders contribute to the collective decision-making process. This view of a continuous 'work in progress' approach to ecotourism is held by other authors to allow for iterations to achieve desired outcomes (Hearne & Santos, 2005; Rowat & Engelhart, 2007). Derived from the review of literature is the conceptual framework for this research illustrated in Figure 1.

In Figure 1, antecedents to ecotourism adoption are driven by internal or external factors, as suggested by other scholars (Boyd & Butler, 1996; Clarke, 1997; Hvenegaard, 1994; Weaver & Lawton, 2007). As a means to align with sustainable practices, these studies postulate that ecotourism is then undertaken by governments or institutions in a top-down manner for stakeholder engagement and subsequent action. Such an outcome should not come as a surprise, as the necessary finances, skills and expertise are in the possession of governments or large agencies (Garrod, 2003; Parker & Khare, 2005). Yet, amidst the justified benefits of a top-down ecotourism approach, Medina (2005) highlighted some concerns as to the legitimacy of ecotourism and the sharing of benefits for the diverse stakeholder groups. Building on the existing scope of ecotourism literature, this research seeks to advance our knowledge about the topic in a destination where ecotourism has no precedence.

Methodology

This research employed a case study methodology. A case study is an analysis of a particular unit, which can comprise of individuals, organisations, or particular groups (Baxter & Jack, 2008). Rather than employing case studies to generalise outcomes, several scholars instead emphasise its theoretical underpinnings to uncover knowledge about a particular phenomenon occurring within the setting (Darke, Shanks, & Broadbent, 1998; Flyvbjerg, 2006; Gerring, 2004; Tellis, 1997). In light of its broad application, the use of case studies is

Table 1. Stages of case study methodology.

Stages	Actions
Preliminary	• Formulate theory from ecotourism literature • Identify three case studies • Generate research questions to be answered
Fieldwork and analysis	• Conduct case studies research • Code themes from each case
Conclusion	• Compare themes and ecotourism practices across cases • Conclusion drawing and verification • Modification to theory

widely found within the literature on ecotourism (Ross & Wall, 1999). Given its benefits to understand ecotourism, the research follows other scholars by adopting the case study methodology to investigate ecotourism in Singapore.

Some extant literature exists to guide the development of a case study methodology. In this space, the works of Eisenhardt (1989) and Yin (1994) have been well-received and cited across various contexts and areas of study (Chetty, 1996; Gibbert, Ruigrok, & Wicki, 2008). The difference between the two approaches to case study methodology lie in the schema to generate theoretical contributions. Eisenhardt (1989) prescribes an inductive approach to case studies and suggests that four to ten cases are optimal numbers to generate theory. In contrast, Yin (1994) approaches case studies to test theoretical underpinnings. As such, he advocates that fewer cases can be justified when existing theory about the topic exists, and where the cases can illuminate more conceptualisations surrounding a topic (Meyer, 2001). Relating to this research, Yin's (1994) approach to case study appears better suited to the needs of understanding ecotourism in Singapore because there is sufficient scope within the existing literature to guide the investigation. Hence, the research follows Yin's (1994) stages of case study research design.

According to Yin (1994), three stages of case study research design are to be considered. The three stages are a preliminary stage, fieldwork and analysis stage and conclusion stage (Noor, 2008). Guided by Yin's (1994) case study methodology, Table 1 illustrates how the three respective stages have been devised.

In the context of Singapore, three case studies were identified to be best suited for the investigation of ecotourism developments. These cases were selected because they have been featured repeatedly in the local media and have only commenced planning with the last 12 months. The three cases featured in this research are:

(1) Sisters' Island Marine Park (https://www.nparks.gov.sg/gardens-parks-and-nature/parks-and-nature-reserves/sisters-islands-marine-park)
(2) Kampung Kampus (http://www.groundupinitiative.org/kampung-kampus)
(3) Kallang Eco-Park (http://news.asiaone.com/news/singapore/eco-village-take-root-kallang-park)

Having identified the three cases for investigation, the next section will provide a brief background of each case how the research has been conducted.

Case 1: Sisters' Island Marine Park

SIMP has a zone of almost 40 hectares to the southern shores of the main island of Singapore. Cornelius (2000) reported that the islands received their names from a legend that is told of two sisters who were separated by a storm following a tragic love story. Despite its sad past, SIMP came to prominence after it became the first marine park in Singapore in July 2014 (Cai, 2014). Since it was established, SIMP has catered to marine enthusiasts by hosting one or two guided walks a month during the low tide to one of the islands – Big Sister's Island. This initiative allows visitors to get a glimpse of marine biodiversity around Singapore. Each tour caters to approximately 45 visitors and at present, remains free of charge. Visitors can only participate in these tours during the allotted booking timeslot and register via emails on a first come first served basis. Each tour is often booked two months ahead and visitors who have been successfully placed in the tour will be contacted via email to confirm their attendance and advised of tour details. The researcher participated in one such tour in December 2014 and the following photographs illustrate what occurred during the visit (Figures 2–4).

Each of these tours lasts around 2 hours and visitors are introduced to the various marine lifeforms in proximity to the island. As the photographs illustrate, visitors often congregate around exhibits such as photograph 3 when a 'seeker' ventures into the tidal

Figure 2. Visitors boarding the ferry to SIMP.

Figure 3. Visitors congregating to observe marine life.

Figure 4. Exhibit of a sea cucumber in plastic container.

areas to place marine creatures into plastic containers and labels each exhibit. A 'seeker' is a trained environmental officer who can be seen in photograph 2 in the background wading into the water and locating marine creatures. During the tour, seekers play in instrumental role to educate visitors about each exhibit.

While the visit can awaken one's senses to the rich source of biodiversity, the tranquillity of the island faced some upcoming challenges. As noted by the guide, the weekends can bring other visitors to the island by means of hired private river taxis. At the cost of around US$100 per trip, a river taxi is available for hire that can carry up to 10 visitors to SIMP. On the island, visitors can get to enjoy a leisurely day by the beach and also a barbeque. Incidentally, it is from such occurrences that the visit to SIMP also revealed issues associated with littering and vandalism, as the subsequent photographs highlight (Figures 5 and 6).

Figure 5. Debris on SIMP.

Figure 6. Damaged garbage bin cover.

When probed as to how SIMP would be managed to minimise environmental issues in the future, the guide and seekers highlighted that this was very much a priority area and yet remains work in progress. This was because at the time of visit, SIMP was under the jurisdiction of Sentosa Development Corporation (SDC), rather than the National Parks Board (NPB) of Singapore. As the SDC oversees the development of Sentosa for the primary purpose of fun and enjoyment, the guide emphasised that the NPB would take a more proactive role in steering future developments at the SIMP. At the time of the visit, most studies were conducted to assess the health condition of the reef and marine biodiversity given that the oil tankers and other cruises also share the same body of water leading to the Straits of Malacca and the South China Sea. As Singapore is a busy cargo terminal, it is therefore reasonable for the focus to conserve marine life to achieve a symbiotic relationship aligned with human activities.

Several volunteers also accompanied the visitors on the SIMP trip. During the visit, volunteers were asked as to what their roles were and also how they envisaged the future of SIMP. To this, one such volunteer proudly replied, 'This gives me an opportunity to showcase our marine diversity to other people and demonstrates that Singapore is not just about its tall buildings and a boring place.' Another volunteer added, 'I can join these tours for free as often as I wish by volunteering. Furthermore, I get to meet other like-minded people and hopefully get them to volunteer with me on other green adventures too.' On the return journey, volunteers encouraged all visitors to stay in the loop of discussions and to post their photographs onto the SIMP Facebook site. This social media site is not managed by SDC but instead a volunteer-initiated space to allow further engagement to take place in a less moderated manner. Evidently, the SIMP visit delivered its objectives of creating an awareness of, and educating visitors as to the various marine biodiversity in Singapore. Following the visit to SIMP, the researcher then subsequently investigated the developments surrounding SIMP through document analysis from official websites, media releases and its Facebook page. The outcomes of these developments towards ecotourism will be addressed later in this paper.

Case study 2: Kampung Kampus (KK)

In September 2014, Ground-Up Initiative (GUI), a non-government organisation, announced its intentions to create Kampung Kampus, a 26,000-hectare site to promote

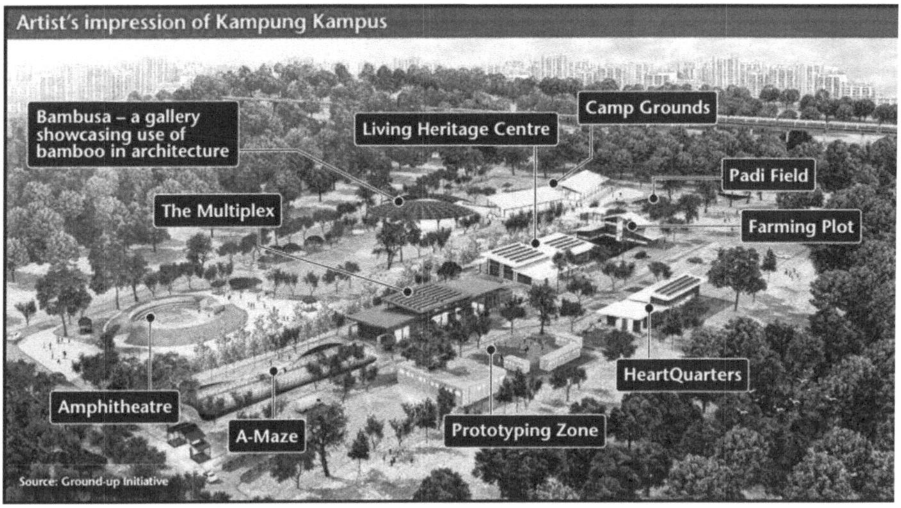

Figure 7. The Kampung Kampus masterplan. Source: Ee (2014).

nature-based education and lifestyles (Phillips, 2014). The name Kampung Kampus literally translates into a village of learning (Ee, 2014). This project was born out of a desire by Mr Tay Lai Hock, also known as 'Chief', to see its visitors gain a sense of community engagement and sustainable living. A masterplan of KK is provided in Figure 7.

The ambitious plans to create KK over the next two decades is lauded my many other observers as a game-changer in the way sustainability is envisioned in Singapore (Liew, 2015). Prior to its inception, the site that KK resides on was a dining zone where cafes and restaurants once thrived. However, following the expiring of this lease, GUI seized upon this opportunity to lobby the government in securing this plot of land for community projects. This successful bid remains one of the few such projects in the country, and as Ee (2014) noted, an inspiration to other Singaporeans who resonated with the overall aims of KK. During the visit to the site in December 2014, the researcher had the opportunity to speak with some volunteers and Mr Tay. When probed as to what the motivations for KK were, Mr Tay explained, '*I gave up my corporate job to do something meaningful. KK turned from a distant dream to a reality and now we have the opportunity to create a new narrative for Singapore.*' This new narrative, as alluded to by Mr Tay, is part of moulding societal beliefs and attitudes towards a 5G generation, where the 5Gs represent graciousness, green, giving, grounded and gratefulness (Panaech, 2014).

In the context of this research, ecotourism is a suitable endeavour of KK given the desire to embrace ecological practices alongside educational messages of sustainability. For this reason, the conversation on developing ecotourism as a proxy to KK's aims and objectives raised much excitement and resulted in a subsequent interview in April 2015 to follow up on what could occur at KK to realise ecotourism potential. While KK is in its early stages, the visit to the premises revealed that there was already in existence small projects geared at sustainable living. The following photographs illustrate the educational messages embedded within various GUI projects (Figures 8 and 9).

Overall, KK epitomises the fundamental building blocks of social networks between GUI and its pool of volunteers. Such an initiative has arguably placed GUI's founder,

Figure 8. A poster at KK about sustainable urban farms (SURF) project.

Mr Tay on a pedestal where he was a nominee for The Straits Times 2014 Singaporean of the Year (Ho, 2015). Clearly, these endeavours have not gone unnoticed and his rallying call for a sustainable Singapore future has found many allies. Through further document analysis of news reports, this research will revisit the role of ecotourism as a catalyst for KK at a later stage.

Case study 3: Kallang Eco-Park (KEP)

The Kallang Eco-Park was the brainchild of Waterways Watch Society (WWS) chairman Mr Eugene Heng, who desired for the adoption of green practices among park users in Singapore (Khew, 2014). KEP is planned for a 400-metre riverside embankment to offer

Figure 9. Photographs on display showing the array of experiences with GUI activities.

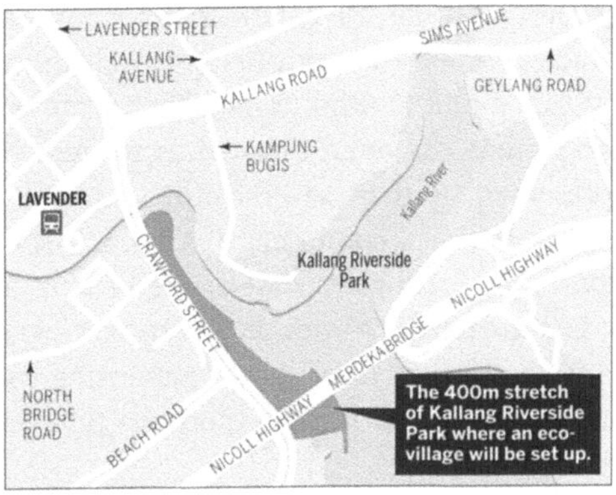

Figure 10. Graphical image of the Kallang Eco-Park. Source: Khew (2014).

visitors an educational precinct about nature and also tours on the Singapore River through kayaks or canoes. Figure 10 outlines the location of KEP in relation to the Singapore River.

Notably, WWS is the first non-government organisation that has successfully secured the management of park premises from the National Parks Board Singapore. The founder of WWS, Mr Heng, retired from a corporate job and fuelled by his passion for environmental quality, started WWS as a movement to bring like-minded volunteers on board (Long, 2013). In recognition of his 17 years contributing to maintaining the river scene in Singapore, Mr Heng was awarded the President's Award for the Environment in 2014 (Sing, 2015). An initial interview with Mr Heng further corroborated his passion, where he added animatedly: 'We live along so many waterways in Singapore, and I find it highly enjoyable to do the things I do. Just the previous weeks, we discovered a family of otters in the river!'

Reflections on the initial research stage

The three case studies exemplified a common point in that the catalyst for proposed ecotourism ventures was derived with almost minimal government involvement. Instead, the desire to undertake ecotourism was driven by the presence of individuals exhibiting a vision of sustainability and championing such a cause. Hence, from the onset, these Singapore case studies illuminate that main discourses conceptualising ecotourism cannot be applied directly as antecedents to ecotourism were propelled out of stakeholder conviction, rather than government action. For this reason, a different trajectory for stakeholder involvement and engagement was required in the research. As such, Figure 11 synthesises how the ecotourism narrative has developed in Singapore.

Figure 11. Form of stakeholder involvement in Singapore ecotourism initiatives.

From Figure 11, the ecotourism narrative detracts from traditional conceptualisations and is very much a characteristic of a bottom-up approach to manufacture ecotourism. This view is justified by Cater (2006) who argued that ecotourism has insofar been a top-down, Western concept that has often dictated as to the why and how ecotourism practices should be represented. Rather, several other scholars rally with a call for a 'bottom-up' approach to construct ecotourism (Farrelly, 2011; Kinnaird & O'Brien, 1996; Sakata & Prideaux, 2013; Southgate, 2006). The benefit of such an approach, as Garrod (2003) suggests, is a greater buy-in occurs from participants that can result in a deeper conviction to bring ecotourism to fruition. Conversing with representatives from each case study showed their desire to further ramp up their efforts and align further initiatives towards ecotourism practices.

Stage 2 – operationalising ecotourism

Building on the interest shown by each of the case studies, the researcher returned to Australia to bring other interested parties to the table in charting an ecotourism blueprint for Singapore. Specifically, two main parties were invited to participate in this project. One party comprised of staff in a research centre focused on sustainability and with experience in ecotourism planning and operations. The other party was Ecotourism Australia (EA), who act as the national body in advising operators on relevant matters. Collectively, both parties were briefed as to what the current state of ecotourism was in Singapore and then to assess if there was interest to collaborate in the project.

The first group consisting of research staff were generally surprised at the rate of nature-based developments in land-scarce Singapore. Nonetheless, they were happy to support initiatives such as training assistance and also conducting field trips to the different sites. However, some concerns were raised, in particular to the SIMP. This was because in the context of Australia and many other parts of the world, the term 'Marine Park' is dedicated for use only under very specific guidelines in terms of marine conservation (Kenchington & Crawford, 1993; McNeill, 1994). Some within the audience also cringed when they saw the exhibits of visitors handling marine creatures during the tour with their bare hands. In tourism literature, contact with plants and animals has often been viewed with detrimental outcomes (Curtin, 2006; Orams, 2002; Uyarra & Cote, 2007). For this reason, the collective view was that the awareness of ecotourism principles and how to manage these natural resources was of utmost importance to assist these initiatives in Singapore.

Ecotourism Australia (EA) were also approached and briefed in terms of the background as to the developments in Singapore. The decision to invite EA was because the organisation is perceived to be highly credible and experienced to guide such initiatives, as Buckley (2004) established. Furthermore, EA is an independent party without any government agenda, and thereby viewed to be an impartial body to undertake objective-based ecotourism (Haaland & Aas, 2010). One of the strategic thrusts of EA is to assess operators for ecotourism certification. Whilst not a compulsory initiative, ecotourism certification is aimed at delivering a consistent level of standards adhered to by practitioners around the country delivering ecotourism experiences (Font, Sanabria, & Skinner, 2003). Obtaining ecotourism certification thereby acts as an endorsement of the rigour associated with the practices of ecotourism operators, which has economic and sustainable benefits

(Jamal et al., 2006; Medina, 2005). The researcher also found that there was an intention for EA to gain traction into other practices, in particular the Asia-Pacific region where ecotourism has shown significant process in the last decade. As such, the team at EA considered that a customised ecotourism certification programme could work and is perhaps necessary given the different political and environmental contexts in Singapore. This would therefore require training a pool of assessors to audit the proposed case study sites. This point was noted and brought to the attention of the case studies.

Stage 3 – re-entry to the scene

The researcher subsequently conducted online meetings with the respective case studies to facilitate subsequent developments for ecotourism. One positive outcome from the initial engagement was the common understanding of ecotourism principles and the desire to weave appropriate ecotourism guidelines to shape future initiatives in light of certification requirements. After all, studies have shown that once tourism developments have commenced and operationalised, greater resistance is felt to modify existing operations to meet ecotourism certification standards (Font, 2002; Zhong, Buckley, & Xie, 2007). The enthusiasm offered by other stakeholders to help with the respective case studies was an encouraging note, especially when each case study faced different challenges. For instance, developing water-based activities at the KEP faced some roadblocks when WWS were interested to launch a remote-operated semi-submersible to assess water quality and the marine diversity in the Singapore River. Mr Heng commented, 'We had to get access from several agencies in order to even release the boat from shore.'

Other issues related to the rolling out of ecotourism include finances. For instance, WWS reportedly had one of their sponsors withdraw due to lack of funds and are actively seeking other stakeholders to help with signage matters (WWS, 2015). As earlier discussed, the lack of funds is a key agenda that will likely determine if ecotourism gets going on the scene in Singapore. To address this issue, GUI has a range of funding initiatives to help develop the KK project. These include corporate teambuilding exercises, inbound student tours and crowdfunding (Wee, 2014). One novel approach to the crowdfunding initiative was to get investors to make an eco-brick (Tay, 2015). This activity not only provided an element of fun and learning, but also symbolised the collective efforts of participants in building a sustainable future for Singapore. As at July 2015, this crowdfunding project raises over US$6,000 towards the implementation of KK (Crowdtivate, 2015). Whilst the amount raised is a far distance to the desired target of US$20,000, Mr Tay felt that the crowdfunding approach was to some extent a success as it had shifted potential stakeholders out of their comfort zone and take active steps towards realising KK. Therefore, Mr Tay revealed that 'Even if KK takes a slightly longer time to become a reality, it would have already left its legacy on the generations to come.'

In the case of SIMP, its planners adopt a more educational approach to gain stakeholder interest and generate involvement. One such avenue was to feature along other related activities in public events, such an annual Festival of Biodiversity (R. Tan, 2015). A significant announcement at this year's edition of the festival was the launch of two dive trails at SIMP commencing in September 2015 (A. Tan, 2015). Evidently, the efforts in harnessing the potential of SIMP as a business model has been expedited, even before ecotourism is put to practice. Another development from the festival was a newly minted Marine

Conservation Action Plan to help manage the coastal habitats in Singapore (Othman, 2015). While specific action steps are currently being formulated, such a plan is likely to further encompass greater stakeholder inputs to ensure that the plan represents a consultative approach towards future endeavours related to SIMP. Funding, in this instance, would appear focused on sourcing external partners, such as banks and research institutions to provide the impetus to manage the dive sites.

In relation to ecotourism, it is also envisaged that these case studies also solicit other stakeholders to get further assistance with planning and certification. All representatives are of the opinion that certification is highly valuable, though the process to obtain certification is very much dependent on time and resources. After all, a common denominator among the case studies is the small full-time workforce and a large pool of volunteers. Most of these individuals have very little background in ecotourism, or even tourism policy and planning. For this reason, the researcher has kept an ongoing engagement with these case studies through regular exchanges of emails, online meetings and also social media communities, such as Facebook.

Discussion and conclusion

Synthesising the outcomes of the research has cumulated with a refined conceptualisation as to what form ecotourism should take in its implementation within Singapore. Whilst literature has guided the understanding of principles and the direction to operationalise ecotourism, the research suggests that following such notions may not always work, and in some cases backfire. This is because a bureaucratic approach to ecotourism can go insofar as allocating people and resources in position, but have very little impact in driving deep-rooted engagement and a sense of ownership (Horton, 2009; Johnston, 2000; Ross & Wall, 1999). Instead, the research pushes for a repositioning of ecotourism to be framed from a bottom-up, and manufactured perspective. This is an appropriate fit to the stakeholder theory that postulates for organisations to incorporate a consultative and value-based approach to realising mutually agreeable outcomes. While the theory often highlights the tensions occurring between stakeholder individuals or groups, there is consensus that the theory works best in ecotourism practice when opinions and decisions are openly voiced and that participation is born out of a sense of identity and conviction (Garrod, 2003; Lawrence et al., 1997; Rowat & Engelhart, 2007).

Applied to the case of Singapore, such an approach is a refreshing change in a country that is often perceived to be a highly regulated environment. Across many facets of business and society, Singapore has become a successful model that is the envy of many nations, though its progress appears to be charted out of meticulous planning and regulatory processes (Lam, 2000). Yet, the engagement with the case studies herein have revealed that minimal government intervention has taken place. This is even more surprising because ecotourism is not likely to have any significant economic returns in the short term, which goes against the core foundations of industrialisation and economic gains that have engineered the success story of Singapore, as espoused by Chang (2003). However, as the nation celebrates its 50th year of independence in 2015, there appears to be a shift in the narrative towards one of citizenship identity making, and could probably account for the reasons as to why ecotourism developments appear to be given a free reign to evolve.

Collectively, the decentralised approach to manufacture ecotourism stems from a vision of individuals with a common passion to deliver a sustainable future for Singapore. This trajectory somewhat deviates from extant literature by engaging stakeholders in a bottom-up approach as the practitioners seek to create an awareness and the value of ecotourism as one tool towards sustainability. This has occurred prior to lobbying the government for access to resources such as financial grants, and has taken place in a largely unmoderated manner. While these initiatives have strongly beneficial outcomes, the rollout of subsequent developments towards ecotourism will nonetheless face issues of finances, time, resources and maintaining momentum.

This project is not without its limitations. As only three case studies were utilised, the outcomes of this research may require further validation. Furthermore, the case studies were based solely within the context of Singapore, and will need subsequent studies to assess if similar findings are obtained in other countries or environments. Despite these limitations, the research has outlined what the key ingredients are to undertake ecotourism planning, especially for destinations or operators seeking to manufacture related experiences where there may not be a precedence.

To conclude, the research borrowed the term 'Envisioning Eden' from Salazar's (2010) book of the same title. Salazar (2010) showcased how tourism experiences are shaped by imaginations of the future and recollections of the past. In a similar trajectory, this research has demonstrated that such a term holds true for the 'eco-warriors' in Singapore. The pursuit of ecotourism is highly congruent to the strategic initiatives of each case study as they envision their own sustainable futures. While these are early days to assess any success, the case studies have already left important blueprints in charting the future of ecotourism in Singapore.

Disclosure statement

No potential conflict of interest was reported by the author.

ORCID

Aaron Tham http://orcid.org/0000-0003-1408-392X

References

Baxter, P., & Jack, S. (2008). Qualitative case study methodology: Study design and implementation for novice researchers. *The Qualitative Report*, *13*(4), 544–559.

Bjork, P. (2000). Ecotourism from a conceptual perspective, an extended definition of a unique tourism form. *International Journal of Tourism Research*, *2*(3), 189–202.

Blamey, R. K. (1997). Ecotourism: The search for an operational definition. *Journal of Sustainable Tourism*, *5*(2), 109–130.

Boyd, S. W., & Butler, R. W. (1996). Managing ecotourism: An opportunity spectrum approach. *Tourism Management*, *17*(8), 557–566.

Brondo, K. V., & Woods, L. (2007). Garifuna land rights and ecotourism as economic development in Honduras' Cayos Cochinos marine protected area. *Ecological and Environmental Anthropology*, *3*(1), 2–18.

Buckley, R. (2004). Partnerships in ecotourism: Australian political frameworks. *International Journal of Tourism Research*, *6*(2), 75–83.

Buckley, R. (2012). Sustainable tourism: Research and reality. *Annals of Tourism Research, 39*(2), 528–546.

Caffyn, A., & Jobbins, G. (2003). Governance capacity and stakeholder interactions in the development and management of coastal tourism: Examples of Morocco and Tunisia. *Journal of Sustainable Tourism, 11*(2–3), 224–245.

Cai, W. J. (2014). *Singapore to get first Marine Park.* Retrieved from http://news.asiaone.com/news/singapore/singapore-get-first-marine-park

Carrier, J. G., & Macleod, D. V. L. (2005). Bursting the bubble: The socio-cultural context of ecotourism. *Journal of the Royal Anthropological Institute, 11*(2), 315–334.

Cater, E. (2006). Ecotourism as a western construct. *Journal of Ecotourism, 5*(1–2), 23–39.

Chan, J. K. L., & Baum, T. (2007). Ecotourists' perception of ecotourism experience in lower Kinabatangan, Sabah, Malaysia. *Journal of Sustainable Tourism, 15*(5), 574–590.

Chang, T. C. (1999). Local uniqueness in the global village: Heritage tourism in Singapore. *The Professional Geographer, 51*(1), 91–103.

Chang, J. H. (2003). Culture, state and economic development in Singapore. *Journal of Contemporary Asia, 33*(1), 85–105.

Chetty, S. (1996). The case study method for research in small- and medium-sized firms. *International Small Business Journal, 15*(1), 73–85.

Clarke, J. (1997). A framework of approaches to sustainable tourism. *Journal of Sustainable Tourism, 5*(3), 224–233.

Cornelius, V. (2000). *Sisters' islands.* Retrieved from http://eresources.nlb.gov.sg/infopedia/articles/SIP_185_2005-01-20.html

Crowdtivate. (2015). *Kampung Kampus: Make a brick!* Retrieved from https://www.crowdtivate.com/projects/view/4610

Curtin, S. (2006). Swimming with dolphins: A phenomenological exploration of tourist recollections. *International Journal of Tourism Research, 8*(6), 301–315.

Darke, P., Shanks, G., & Broadbent, M. (1998). Successfully completing case study research: Combining rigour, relevance and pragmatism. *Information Systems Journal, 8*(4), 273–289.

de Araujo, L. M., & Bramwell, B. (1999). Stakeholder assessment and collaborative tourism planning: The case of Brazil's costa Dourada project. *Journal of Sustainable Tourism, 7*(3–4), 356–378.

Donohoe, H. M., & Needham, R. D. (2006). Ecotourism: The evolving contemporary definition. *Journal of Ecotourism, 5*(3), 192–210.

Ee, D. (2014). *Kampung Kampus: Eco-friendly site to take root in Khatib.* Retrieved from http://ifonlysingaporeans.blogspot.com.au/2014/11/kampung-kampus-eco-friendly-site-to.html

Eisenhardt, K. M. (1989). Building theories from case study research. *Academy of Management Review, 14*(4), 532–550.

Farrelly, T. A. (2011). Indigenous and democratic decision-making: Issues from community-based ecotourism in the Boumā National Heritage Park, Fiji. *Journal of Sustainable Tourism, 19*(7), 817–835.

Fennell, D. A. (2001). A content analysis of ecotourism definitions. *Current Issues in Tourism, 4*(5), 403–421.

Fletcher, R. (2009). Ecotourism discourse: Challenging the stakeholder theory. *Journal of Ecotourism, 8*(3), 269–285.

Flyvbjerg, B. (2006). Five misunderstandings about case-study research. *Qualitative Inquiry, 12*(2), 219–245.

Font, X. (2002). Environmental certification in tourism and hospitality: Progress, process and prospects. *Tourism Management, 23*(2), 197–205.

Font, X., Sanabria, R., & Skinner, E. (2003). Sustainable tourism and ecotourism certification: Raising standards and benefits. *Journal of Ecotourism, 2*(3), 213–218.

Galley, G., & Clifton, J. (2004). The motivational and demographic characteristics of research ecotourists: Operation Wallcea volunteers in southeast Sulawesi, Indonesia. *Journal of Ecotourism, 3* (1), 69–82.

Garrod, B. (2003). Local participation in the planning and management of ecotourism: A revised model approach. *Journal of Ecotourism, 2*(1), 33–53.

Gerring, J. (2004). What is a case study and what is it good for? *American Political Science Review, 98*(2), 341–354.
Gibbert, M., Ruigrok, W., & Wicki, B. (2008). What passes as a rigorous case study? *Strategic Management Journal, 29*(13), 1465–1474.
Gladstone, W., Curley, B., & Shokri, M. R. (2013). Environmental impacts of tourism in the gulf and the Red Sea. *Marine Pollution Bulletin, 72*(2), 375–388.
Gray, N. J., & Campbell, L. M. (2007). A decommodified experience? Exploring aesthetic, economic and ethical values for volunteer ecotourism in Costa Rica. *Journal of Sustainable Tourism, 15*(5), 463–482.
Gui, Y., Fang, Y., & Liu, J. (2004). Community-based ecotourism in nature reserve of China. *Chinese Geographical Science, 14*(3), 276–282.
Gurung, D. B., & Scholz, R. W. (2008). Community-based ecotourism in Bhutan: Expert evaluation of stakeholder-based scenarios. *International Journal of Sustainable Development & World Ecology, 15*(5), 397–411.
Gylfason, T., & Zoega, G. (2006). Natural resources and economic growth: The role of investment. *The World Economy, 29*(8), 1091–1115.
Haaland, H., & Aas, O. (2010). Eco-tourism certification – does it make a difference? A comparison of systems from Australia, Costa Rica and Sweden. *Scandinavian Journal of Hospitality and Tourism, 10*(3), 375–385.
Hearne, R. R., & Santos, C. A. (2005). Tourists' and locals' preferences toward ecotourism development in the Maya biosphere reserve, Guatemala. *Environment, Development and Sustainability, 7*(3), 303–318.
Henderson, J. (2004). Food as a tourism resource: A view from Singapore. *Tourism Recreation Research, 29*(3), 69–74.
Henderson, J., Foo, K., Lim, H., & Yip, S. (2010). Sports events and tourism: The Singapore formula one grand prix. *International Journal of Event and Festival Management, 1*(1), 60–73.
Higham, J., & Luck, M. (2002). Urban ecotourism: A contradiction in terms? *Journal of Ecotourism, 1*(1), 36–51.
Ho, O. (2015). *Inspiration to us all: He helps others 'put down roots'*. Retrieved from http://www.straitstimes.com/singapore/inspiration-to-us-all-he-helps-others-put-down-roots
Horton, L. R. (2009). Buying up nature: Economic and social impacts of Costa Rica's ecotourism boom. *Latin American Perspectives, 36*(3), 93–107.
Hui, T. K., Wan, D., & Ho, A. (2007). Tourists' satisfaction, recommendation and revisiting Singapore. *Tourism Management, 28*(4), 965–975.
Hvenegaard, G. T. (1994). Ecotourism: A status report and conceptual framework. *Journal of Tourism Studies, 5*(2), 24–35.
Jackson, S. (2007). Attitudes towards the environment and ecotourism of stakeholders in the UK tourism industry with particular reference to ornithological tour operators. *Journal of Ecotourism, 6*(1), 34–66.
Jamal, T., Borges, M., & Stronza, A. (2006). The institutionalization of ecotourism: Certification, cultural equity and praxis. *Journal of Ecotourism, 5*(3), 145–175.
Johnston, A. (2000). Indigenous peoples and ecotourism: Bringing indigenous knowledge and rights into the sustainability equation. *Tourism Recreation Research, 25*(2), 89–96.
Kenchington, R., & Crawford, D. (1993). On the meaning of integration in coastal zone management. *Ocean & Coastal Management, 21*(1–3), 109–127.
Khew, C. (2014). *Eco-village to take root in Kallang Park*. Retrieved from http://news.asiaone.com/news/singapore/eco-village-take-root-kallang-park
Kinnaird, M. F., & O'Brien, T. G. (1996). Ecotourism in the Tangkoko DuaSudara nature reserve: Opening Pandora' box? *Oryx, 30*(1), 65–73.
Kontogeorgopoulos, N. (2004). Conventional tourism and ecotourism in Phuket, Thailand: Conflicting paradigms or symbiotic partners? *Journal of Ecotourism, 3*(2), 87–108.
Kruger, O. (2005). The role of ecotourism in conservation: Panacea or Pandora's Box? *Biodiversity & Conservation, 14*(3), 579–600.

Lai, P., & Nepal, S. K. (2006). Local perspectives of ecotourism development in Tawushan nature reserve, Taiwan. *Tourism Management, 27*(6), 1117–1129.

Lam, N. M. K. (2000). Government intervention in the economy: A comparative analysis of Singapore and Hong Kong. *Public Administration and Development, 20*(5), 397–421.

Lawrence, T. B., Wickins, D., & Phillips, N. (1997). Managing legitimacy in ecotourism. *Tourism Management, 18*(5), 307–316.

Lee, C. G. (2010). Health care and tourism: Evidence from Singapore. *Tourism Management, 31*(4), 486–488.

Leng, C. H. (2010). Medical tourism and the state in Malaysia and Singapore. *Global Social Policy, 10*(3), 336–357.

Lew, A. (1992). Place representation in tourist guidebooks: An example of Singapore. *Singapore Journal of Tropical Geography, 12*(2), 124–137.

Liew, C. (2015). Building a future from ground up with Tay Lai hock: Back to Kampung. *Citygreen, 10*, 124–129.

Litvin, S., & Chiam, M. K. M. (2014). Ecotourism – a study of purchase proclivity, revisited. *Journal of Vacation Marketing, 20*(2), 175–180.

Liu, X., & Bao, J. (2004). Unravelling ecotourism practice: Problem analysis based on stakeholders. *Chinese Geographical Science, 14*(1), 82–89.

Lohmann, G., Albers, S., Koch, B., & Pavlovich, K. (2009). From hub to tourist destination –an explorative study of Singapore and Dubai's aviation-based transformation. *Journal of Air Transport Management, 15*(5), 205–211.

Long, S. (2013). *Eugene Heng is watchdog of Singapore waterways*. Retrieved from http://news.asiaone.com/news/singapore/eugene-heng-watchdog-singapore-waterways

McNeill, S. E. (1994). The selection and design of marine protected areas: Australia as a case study. *Biodiversity & Conservation, 3*(7), 586–605.

Medina, L. K. (2005). Ecotourism and certification: Confronting the principles and pragmatics of socially responsible tourism. *Journal of Sustainable Tourism, 13*(3), 281–295.

Meyer, C. B. (2001). A case in case study methodology. *Field Methods, 13*(4), 329–352.

Newman, P. (2014). Biophilic urbanism: A case study of Singapore. *Australian Planner, 51*(1), 47–65.

Noor, K. B. M. (2008). Case study: A strategic research methodology. *American Journal of Applied Sciences, 5*(11), 1602–1604.

Ooi, G. L. (2007). Urbanization in Southeast Asia: Assessing policy process and progress toward sustainability. *Journal of Industrial Ecology, 11*(2), 31–42.

Orams, M. B. (1995). Towards a more desirable form of ecotourism. *Tourism Management, 16*(1), 3–8.

Orams, M. B. (2002). Feeding wildlife as a tourism attraction: A review of issues and impacts. *Tourism Management, 23*(3), 281–293.

Othman, L. (2015). *More protection for marine habitats with new conservation plan*. Retrieved from http://www.todayonline.com/singapore/more-protection-marine-habitats-new-conservation-plan

Panaech, S. (2014). *Nee Soon GRC to host $6m 21st century Kampung culture community*. Retrieved from http://news.asiaone.com/news/singapore/nee-soon-grc-host-6m-21st-century-kampung-culture-community

Parker, S., & Khare, A. (2005). Understanding success factors for ensuring sustainability in ecotourism development in Southern Africa. *Journal of Ecotourism, 4*(1), 32–46.

Phillips, D. (2014). *'Balik Kampung' programme by ground up initiative*. Retrieved from http://thepresentpurpose.com/balik-kampung-programme-by-ground-up-initiative-kampung-kampus-volunteer-and-get-involved/

Preece, R. (2012). *Lighting up the night sky: Giant man-made 'supertrees' illuminate city as part of huge £350million gardens project to make Singapore 'botanical capital of the world'*. Retrieved from http://www.dailymail.co.uk/news/article-2157488/Gardens-By-The-Bay-Supertrees-Singapore-light-night-sky.html

Ross, S., & Wall, G. (1999). Evaluating ecotourism: The case of North Sulawesi, Indonesia. *Tourism Management, 20*(6), 673–682.

Rowat, D., & Engelhart, U. (2007). Seychelles: A case study of community involvement in the development of whale shark ecotourism and its socio-economic impact. *Fisheries Research*, *81*(1), 109–113.

Sakata, H., & Prideaux, B. (2013). An alternative approach to community-based ecotourism: A bottom-up locally initiated non-monetised project in Papua New Guinea. *Journal of Sustainable Tourism*, *21*(6), 880–899.

Salazar, N. B. (2010). *Envisioning Eden: Mobilizing imaginaries in tourism and beyond*. Oxford: Berghahn Books.

Salleh, N. A. M. (2014). *Singaporeans can play a role in retaining Pulau Ubin's charm, says PM Lee*. Retrieved from http://www.straitstimes.com/news/singapore/more-singapore-stories/story/singaporeans-can-play-role-retaining-pulau-ubins-charm-s

Sharpley, R. (2006). Ecotourism: A consumption perspective. *Journal of Ecotourism*, *5*(1–2), 7–22.

Siau, M. (2015). *UNESCO delegates laud Singapore's nomination of Botanic Gardens as World Heritage Site*. Retrieved from http://www.channelnewsasia.com/news/singapore/unesco-delegates-laud/1961592.html

Sing, M. G. (2015). *Turning the tide*. Retrieved from http://sg.asiatatler.com/society/turning-the-tide

Southgate, G. R. J. (2006). Ecotourism in Kenya: The vulnerability of communities. *Journal of Ecotourism*, *5*(1–2), 80–96.

Stronza, A. (2007). The economic promise of ecotourism for conservation. *Journal of Ecotourism*, *6* (3), 210–230.

Tan, A. (2015). *See spore's latest nature parks – under the sea*. Retrieved from http://www.straitstimes.com/opinion/see-spores-latest-nature-parks-under-the-sea

Tan, R. (2015). *Joyful June at the Sisters Islands Marine Park, with a tinge of sadness*. Retrieved from http://sistersislandmarinepark.blogspot.com.au/2015/06/joyful-june-at-sisters-islands-marine.html

Tang, Z. (2015). An integrated approach to evaluating the coupling coordination between tourism and the environment. *Tourism Management*, *46*, 11–19.

Tay, E. (2015). *Presenting #Build5G by ground-up initiative (GUI)*. Retrieved from http://makerfairesingapore.com/2015/07/12/presenting-build5g-by-ground-up-initiative-gui/

Tellis, W. M. (1997). Application of a case study methodology. *The Qualitative Report*, *3*(3), 1–19.

Teo, P., & Huang, S. (1995). Tourism and heritage conservation in Singapore. *Annals of Tourism Research*, *22*(3), 589–615.

Tsai, S. (2012). Place attachment and tourism marketing: Investigating international tourists in Singapore. *International Journal of Tourism Research*, *14*(2), 139–152.

Tsaur, S., Lin, Y., & Lin, J. (2006). Evaluating ecotourism sustainability from the integrated perspective of resource, community and tourism. *Tourism Management*, *27*(4), 640–653.

Uyarra, M. C., & Cote, I. M. (2007). The quest for cryptic creatures: Impacts of species-focused recreational diving on corals. *Biological Conservation*, *136*(1), 77–84.

Waterways Watch Society. (2015). *E-News*. Retrieved from http://www.wws.org.sg/enews.php

Weaver, D. (2005). Comprehensive and minimalist dimensions of ecotourism. *Annals of Tourism Research*, *32*(2), 439–455.

Weaver, D., & Lawton, L. (2007). Twenty years on: The state of contemporary ecotourism research. *Tourism Management*, *28*(5), 1168–1179.

Wee, L. (2014). *Kampung chief*. Retrieved from http://www.c3a.org.sg/Learning_contect.do?id=17558

Wight, P. (1993). Ecotourism: Ethics or eco-sell? *Journal of Travel Research*, *31*(3), 3–9.

Wu, B. (2012). *The promise and challenge of ecotourism*. Retrieved from http://ink.library.smu.edu.sg/cgi/viewcontent.cgi?article=1107&context=lien_research

Yap, W. M. (2007). Bringing casinos to the shores of Singapore. *Gaming Law Review*, *11*(4), 440–448.

Yeung, S., Wong, J., & Ko, E. (2004). Preferred shopping destination: Hong Kong versus Singapore. *International Journal of Tourism Research*, *6*(2), 85–96.

Yin, R. K. (1994). *Case study research: Design and methods*. Beverley Hills, CA: Sage.

Zhong, L., Buckley, R., & Xie, T. (2007). Chinese perspectives on tourism eco-certification. *Annals of Tourism Research*, *34*(3), 808–811.

Authenticating eco-cultural tourism in Kazakhstan: a supply side perspective

Guillaume Tiberghien, Hamish Bremner and Simon Milne

ABSTRACT

The country of Kazakhstan is not a well-known tourism destination, either globally or within Central Asia. Although the number of inbound tourists remains relatively small, Kazakhstan possesses numerous tourism attractions based on its past Soviet times, nomadic culture and a variety of unique landscapes. This paper presents the findings that focus on the perceptions of authenticity of tourism providers involved in the development of two Kazakhstani eco-cultural tours. Empirical research is based on in-depth semi-structured interviews conducted between August 2011 and May 2012 which were carried out with community members, policymakers and tourism developers. Using a qualitative case study research approach, the analysis of data identifies tourism providers' authentication positions on various themes and indicators of authenticity for Kazakhstani eco-cultural tourism. Several topic areas including the geographical imagination (nomadic cultural landscapes), crafts purchased by tourists (nomadic ethnic art), and performative spaces (nomadic home-stays and nomadic food) are identified as sources of authentic tourism experiences for visitors. For each topic area, various stakeholders' perceptions of authenticity are presented. The article examines various aspects of the commodification of nomadic culture in a post-Soviet heritage and details the role authenticity plays in the planning and development of Kazakhstani tourism and local community participation and empowerment.

Introduction

Questions regarding authenticity are central to literature in cultural heritage and tourism development (Cohen, 1988; Cohen & Cohen, 2012; Jamal & Hill, 2004; Wang, 1999; Xie, 2011). How tourism providers can shape the reality for tourists and provide a depiction of the social and economic situation in the destination is becoming increasingly important for 'off the beaten track' tourism destinations who seek to attract visitors. Jamal and Hill (2004) argue that it is critical to place this 'authenticity' in the context of how and why the tourism industry defines and presents its version of the genuinely local in both ecological and cultural aspects of the tourism experience. For Xie (2011, p. 32), 'the significance of the debate of authentic culture depends on what Bhabha (2004) called an "active agency", where cultural meaning should be negotiated by a variety of parties.' Xie (2011, p. 37) further points out that, 'it is thus important to shift the direction of research from authenticity to *authentication* and identify the positions of the stakeholders who authenticate tourism and its resources.' As heritage, and the presentation of heritage, is intrinsically a contested phenomenon subject to various stakeholders' interpretations, the concept of authenticity applied to Kazakhstani cultural heritage has become particularly relevant to tourism development in the country.

Several studies have addressed the question of authenticity and the development of tourism in Asia (Buckley, Ollenburg, & Zhong, 2008; Kolas, 2008; Suntikul, Butler, & Airey, 2010; Werner, 2003; Xie, 2001, 2011; Xie & Wall, 2002; Yang & Wall, 2009), but the question of authenticity regarding Kazakhstani eco-cultural tourism practices in a post-Soviet era remains unexplored. In particular, there is a need for investigation regarding how the notion of authenticity is understood, perceived and constructed by multiple stakeholders involved in the development of eco-cultural tourism in Kazakhstan; and in what ways a deeper understanding of authenticity can be used to inform the planning and future development of Kazakhstani tourism.

This article will first review the literature pertaining to authenticity and stakeholder involvement in the development of sustainable tourism. This is followed by a description of the case studies involved in the research, before outlining the research approach for the study. The discussion then examines various aspects of the commodification of Kazakhstani nomadic culture in a post-Soviet era and discusses the different perspectives regarding the authentication of Kazakhstani nomadic culture. Tourism management strategies and practices related to this authenticity are also considered in the discussion.

Stakeholders and dimensions of authenticity

A stakeholder is identified as 'any group or individual who can affect or is affected by tourism development in an area' (Freeman, 1984, p. 46) and must have a legitimate interest in the organisation (Donaldson & Preston, 1995). Using the 'descriptive' aspect of stakeholder theory developed by Donaldson and Preston (1995) for sustainable tourism, Byrd (2007, p. 7) details that:

> Stakeholder theory can describe the multiple elements of tourism in a community, the history of tourism development in the community, the procedures and policies that relate to the development and management of tourism in the area, the types of attractions in the community, the overall economic impact to the community, and the connections between the different agencies and organizations that are involved in tourism.

An understanding of the stakeholders' views in the development of Kazakhstani eco-cultural tourism practices is essential in the construction of visitors' experiences. As Xie (2011, p. 41) advocates, it becomes important to explain 'how authentication is carried out as a particular culture's form of expression by institutionalising and authorising social practices and knowledge'. In the Kazakhstani context, an understanding of stakeholders' authentication positions of tourism and its resources is thus warranted.

The supply side of eco-cultural tourism in Kazakhstan comprises a range of different stakeholders including community members (home-stay providers), policymakers (government officials from the Ministry of Tourism and Sport of the Republic of Kazakhstan) and tourism developers including tour operators, local and international non-governmental organisations (NGOs). The difficulty with the concept of authenticity in tourism studies is that it is a criterion used by tourists as observers but 'whether the "tourees" observed by the tourists possess such a concept and, if so, which traits of their culture they consider as authentic, is almost never raised' (Cohen, 1988, p. 374). Cultural arts and performances are also being created for the sake of tourism by the local populations which, according to Cohen (1988) participate in a form of 'cultural continuity' that the host communities totally integrate as a new form of cultural expression. Defining what is authentic – or what is not – is also highly dependent on the political context of the destination. When referring to the landscapes of Mongolia, Buckley et al. (2008, p. 57) argue that constructing tourism products is highly dependent on a local political willingness as 'constructing tourism products based on indigenous populations' cultural landscapes may become one way for these peoples to reaffirm their own territorial and cultural identities, either for internal or for external political reasons'.

For Wang (1999, p. 350), 'authenticity is relevant to some kinds of tourism such as ethnic, history or culture tourism, which involve the representation of the Other or of the past.' Kirshenblatt-Gimblett (1998) describes three types of time involved in heritage tourism: 'historic time', 'heritage time' and 'visitor time'. These three categories help to situate the three dimensions of authenticity (objective, constructed and existential) from Jamal and Hill (2004, p. 358). Reisinger and Steiner (2006) and Wang (1999) state that objectivist and modernist authors argue that there is an evident, objective basis for judging authenticity. Historic time refers to the time when an object or event has been evaluated for its authenticity, typically assessed by historians, scientists or archaeologists who are evaluating the time, date and location as markers of authentication of objects of interests. Conversely, constructivists suggest that tourists' experiences or perceptions can be authentic even when they are perfectly aware that the setting has been contrived (Cohen, 1988). Heritage time refers to the constructivist approach and relates to the meaning visitors can give to the object or event being evaluated in terms of its authenticity. At this stage, the role of both the public and private sectors shapes the heritage story and narratives of the place that are interpreted and consumed by the visitors. Some scholars have argued for a more existential approach to the question of authenticity (Hughes, 1995; Wang, 1999), with individuals seen as creating a sense of truth within themselves (MacLeod, 2006). For Reisinger and Steiner (2006) and Wang (1999), existential authenticity is not object based but activity based and can be divided into two dimensions: intra-personal (bodily feelings) and inter-personal (self-making). For Jamal and Hill (2004, p. 357), the visitor time can be associated with

> A transcendence of time, when the tourist is aware that an event took place in another time, but is also aware of that moment's importance in relation to the tourist's own life, so that the experiential moment can be simultaneously in the past, present and even future.

In this study, the examination of theoretical foundations of the concept of authenticity in cultural-heritage tourism assist in the task of defining epistemological approaches that frame the construction of authenticity in Kazakhstani eco-cultural tourism practices, and help uncovering multi-dimensional indicators of authenticity for heritage and cultural tourism management in Kazakhstan that are presented later in the paper.

Research context

Although the number of inbound tourists remains relatively small, Kazakhstan possesses numerous tourism attractions based on its past Soviet times, nomadic culture and a variety of unique landscapes. Since the independence of Kazakhstan in 1991, the demand for cultural heritage experiences is increasingly leading to a number of new nature and cultural-tourism products providing local communities' perspectives in the form of 'eco-cultural' tourism, a form of tourism presented by Wallace and Russell (2004) as a concept in which ecological and cultural aspects of a landscape are combined to create experiences for tourists. Eco-cultural tourism projects have been developed in Kazakhstan by a range of local tourism providers and promoted by national and international NGOs including the German Nature and Biodiversity Conservation Union (NABU), the Kazakh NGO 'Eco-museum Karaganda' and the governmental 'Kazakhstan Tourism Association' (KTA). Eco-sites and eco-tours with a network of home-stays were developed in Central Kazakhstan and managed by the operator, 'Nomadic Travel Kazakhstan'. One of the eco-tours and first case study of this research, the three-day Kyzylarai tour, combines different eco-cultural elements such as visits to the granite sepulchres of Begazy as well as rock paintings dating from the Bronze Age. Visitors are accommodated in home-stays in the village of Shabanbai Bi where local population have preserved the skill to produce handmade fur products and nomadic food specialities including traditional dishes made out of horse meat (*bes barmak*), local horse and camel fermented-milk products (*kymiz* and *shubat*), which are made available to the visitors. The second case study, the three-day Tulip tour was developed in the Southern part of the country by independent tour operators including a member of KTA where visitors are accommodated in yurts specifically constructed next to the Kanshengel village. Tourists are offered the opportunity to visit archaeological sites from the Bronze Age, and an exploration of the steppes and associated flora and fauna. A visit to a camel farm where visitors are offered the opportunity to taste *shubat* (camel milk) and derived camel milk products (*kurt*) from traditional nomadic culture is also included. The tour is designed for visitors who want an experience in sleeping in a yurt. In order to keep a certain level of comfort, home-stay providers offered visitors the choice of sleeping in beds or on *körpes* (traditional mattresses on the floor), as well as proper sanitary conditions and toilets.

Methodology

Rickly-Boyd (2012, p. 274) argues that 'while there are clearly multiple ontological and epistemological perspectives of authenticity, few researchers use only one paradigm.' Rickly-Boyd discusses that when analysing the authenticity of a tourism experience, one has to consider the strong interaction between object, site and experience, which are not mutually exclusive. Relativist ontology assumes that realities exist in the form of multiple mental constructions, socially and experientially based, local and specific, dependent for their form and content on the persons who hold them (Guba & Lincoln, 1994; Pernecky & Jamal, 2010). Because the study also includes 'tourists and residents who engage in sense-making, narrative and interpretive meaning-making encounters with situated place and contextual space' (Jamal & Hill, 2004, p. 21), this research is predominantly situated in a 'constructivist/interpretivist thought and practice' (Hollinshead, 2006, p. 43) that is grounded in an 'essentially relativist' ontology (Chambers, 2007, p. 109) and a subjectivist epistemology (the research views reality as subjective and constructed by the individuals involved in the research process) (Creswell, 2007; Denzin & Lincoln, 2000).

In regards to the different epistemological and philosophical positions that have a stake in the conceptualisation of authenticity, the study adopted a constructivist paradigm as a managerially more adequate position. In particular, the constructivist paradigm was appropriate for uncovering the managerial implications of the process of commodification of Kazakhstani cultural heritage. The purpose of this study is not only to describe what was experienced by various tourism providers but also to understand how their perception of authenticity of the Kazakhstani Soviet and post-Soviet heritage is understood and constructed as a factor influencing the development of eco-cultural tourism in the country. The researchers thus elaborated around a central phenomenon which included the concept of authenticity in tourism studies, causal conditions, strategies for eco-tourism development, the Kazakhstani socio-cultural context and consequences for the local tourism industry.

Phillimore and Goodson (2004, p. 3) argue that qualitative methods are employed 'to collect data about activities, events, occurrences and behaviours and to seek an understanding of actions, problems and processes in their social context.' Tourism knowledge gathering has been generally characterised by case studies, area-specific discussions, examples of best practices, and one-off or one-time research (Hall, Williams, & Lew, 2004). Case studies seek to undertake in-depth analysis of single cases or multiple cases and develop themes, assertions and explanations specific to the cases, which are bounded by time and place (Creswell, 1998). Case studies provided the major source of empirical evidence for the analysis of the question of authenticity in Kazakhstani eco-cultural tourism practices. The evidence from multiple cases is often considered more compelling than using a single case, and the overall study is therefore regarded as being more robust (Yin, 2009). More importantly, the analytic benefits from having two cases may be substantial (Yin, 2009, p. 61). After one of the researcher met with various stakeholders (community members, policymaker and tourism developers) involved in Kazakhstani eco-cultural tourism in the country during the first international ecotourism conference held in Central Kazakhstan in August 2010, a multiple-stakeholder approach was chosen to understand the development of Kazakhstani eco-cultural tourism. During the conference, the Kyzylarai tour in Central Kazakhstan and the Tulip tour in the southern part of the country were chosen as sources of empirical evidence for the study. Both tours included archaeological sites from the Bronze Age and encompassed various aspects of the remains of nomadic culture heritage, including travelling in the landscape of the steppes, eating local food with home-stay providers and being offered to buy local craft-making. Both case studies represent key eco-cultural tourism practices but offered contrasting situations in terms of research setting and organisation of the tour.

Due to the recent development of the Kyzylarai and Tulip tours (launched in July 2010 and May 2012, respectively), the researchers had limited available information about the population from which the sample would be taken. Therefore, purposive or judgmental sampling method was used in order to select unique cases that were especially informative about the development of eco-cultural tourism projects in Kazakhstan; this sampling method is preferred in situations when an expert uses judgment in selecting cases with a specific purpose in mind (Neuman, 2009). The multi-stakeholder approach allowed the researcher to interview all the different groups of populations involved in the development of the Kyzylarai and Tulip tours. After a phase of data collection in Central Kazakhstan with the Kyzylarai tour in August 2011, the researcher embarked on the Tulip tour in South Kazakhstan in May 2012; collecting data from a second tour expanded the sample of stakeholders. Nineteen semi-structured in-depth interviews using open-ended questions were conducted between August 2011 and May 2012 with various stakeholders who were directly or indirectly involved with the development of the Kyzylarai and Tulip tours including local government officials, tourism operators, home-stay providers and NGO coordinators. Notes were taken throughout the one hour long interviews and reviewed to draw out key themes.

An analytical mode of generalisation was chosen due to the ability to enrich the results about the variation of the perception of authenticity in eco-cultural tours. Yin (2009) argues that the examination of word tables from cross-case patterns strongly relies on argumentative interpretation. The interview transcripts were coded line by line, looking for recurrent themes. Field notes, interview transcripts, and the concurrent integration of secondary interdisciplinary literature developed and refined the emergent themes. Analyses and interpretations were contextualised by describing the various stakeholders being interviewed, by explaining the Kazakhstani context, and by highlighting the infancy stage of eco-cultural tourism development in the country. Some questions used in semi-structured interviews ('What is your definition of an authentic tourism experience?', 'How would you characterise contemporary nomadic culture in Kazakhstan?') involved rich and complex answers from the participants. Multiple iterations of coding were used to confirm the validity of the data analysis. Following this approach, the researchers managed with complementary word tables to draw cross-case patterns about different stakeholders' perception of authenticity regarding eco-cultural tourism in Kazakhstan. Finally, themes were continuously compared and checked against the empirical material in order to make the findings and conclusions credible.

Research findings

Several themes including the geographical imagination (nomadic cultural landscapes), crafts purchased by tourists (nomadic ethnic art), and performative spaces (nomadic home-stays and nomadic food) are highlighted by tourism providers as sources of authentic tourism experiences for visitors. For each theme, various stakeholders' authentication positions are now presented.

Nomadic cultural landscapes

In their study about cultural landscapes in Mongolian tourism, Buckley et al. (2008, p. 48) define the term cultural landscape as being intricately entwined with the populations who inhabit them:

> A cultural landscape is an area where the landforms have been created by human culture as well as by nature; human culture has been created by the landscapes as well as the people; and each now depends upon and continues to exist because of the other.

Cultural landscapes (steppes) and the different meanings and symbolic aspects of the Kazakhstani flora and fauna are seen to be essential elements to portraying an authentic tourism experience. One of the NGO coordinators emphasised the need to 'combine natural sights with ethnical, historical and cultural features'. According to the employees of 'Nomadic Travel Kazakhstan', the flora and fauna of the steppes are considered as key elements in portraying an authentic image of nomadic culture. This view is shaped by local home-stay providers, for whom their association with the steppes includes 'wild nature', 'unspoilt and unique landscapes', 'natural sightseeing' and 'diversity of deserts, mountains and pine forests'. Archaeological sites encountered during the tours (ancient stone carvings, or petroglyphs) are perceived by operators of both tours to have a 'true authentic meaning' and are highlighted as a cultural component in the eco-tours. The archaeological site of Begazy, which is included in the Kyzylarai tour, offers a unique opportunity for visitors to witness an ancient 'authentic' historical site from the Bronze Age. Similarly, the Tulip tour encompasses the UNESCO World Heritage Site of Tamgaly, where petroglyphs from the Bronze Age can be found. These sites are acknowledged by all tourism stakeholders as depicting an 'objective' ancient nomadic way of life for visitors, or the visitors concept 'historic time' as described by Kirshenblatt-Gimblett (1998).

The Kazakhstani landscapes provide a 'feeling of being a nomad' and are a symbolic depiction of an authentic element to the visitor's tourism experience. The transcendence of time associated with the visitors' gaze of the steppes refers to what Wang (1999, p. 355) calls 'the authenticity of the origins felt by the visitors and correspondingly to the authenticity projected onto toured objects by tourists and tourism producers in terms of their imagery, expectations, preferences and beliefs.' One of the home-stay providers from the village of Shabanbai Bi specifically emphasised the mystery of the inhabited steppes for some international visitors, as 'upon arrival, visitors have romantic and idealised views of traditional Kazakhstani nomadic culture, typically steppes landscapes and mobile dwellings (yurt-camps).' NGO coordinators suggest that the Kazakhstani cultural landscapes often constitute the main authentic aspect of the visitor experience. From a more pragmatic perspective, NGO coordinators encourage the need to (re)connect Kazakhstani semi-nomadic people and visitors with traditional nomadic lifestyles by using cultural landscapes as an 'additional component in the tourism adventure'. They highlight that, in particular, this connection gives visitors the possibility to travel in cultural landscapes on horseback and experience, upon availability, a yurt nomadic shelter with the local population, and therefore 'meet visitors'' expectations about traditional nomadic culture'.

Nomadic ethnic art

In the village of Shabanbai Bi, most of the home-stay providers are ready to offer handmade fur crafts in the form of *kilims* and *körpes* (carpets) and *tapochkis* (slippers), depending on tourist demand. The workshop on fur production organised by tourism operators and NGO coordinators in the village of Shabanbai Bi highlights the importance of genuine fur craft-making for the local population. Fur craft-making represents the heritage of the material and spiritual culture of the tribes who inhabited the Great Kazakh steppes in ancient times. Its development is closely connected to the nomadic characteristics of the traditional social and economic aspects of the Kazakh society as Kazakh lifestyle was mostly defined by folk craft and trade.

The souvenirs are individually produced by local villagers who sell a range of handcrafted items that are sometimes exclusively designed for the needs of the visitors. Craft-making, according to Littrell, Anderson, and Brown (1993) have indicators of authenticity. This is similar to the Kazakh context whereby external criteria (aesthetics, production techniques or time/place of manufacture) and internal criteria (whether crafts are appealing or useful when they arrive home) are found to be important markers of authenticity. The makers of souvenirs argue some carpets made in the village of Shabanbai Bi are still objectively authentic, incorporating the embroideries and ornaments from ancient times, even if they are now mostly made out of cotton rather than camel wool. Depending on the tourists' demand, home-stay providers in the Shabanbai Bi village offer to make either traditional or new crafts out of fur materials. Thus new crafts in fur materials (carpets), jewellery (rings, earrings) or even toys for children are continuously reinvented for tourism purposes and plays a significant part in the revival of the traditional Kazakh culture despite the fact that the fabrication is made on machines dating from Soviet times.

Performative spaces: nomadic home-stays

The notion of eco-cultural tourism and authenticity is important for the tours because most of the villages in Kazakhstan are largely physically unchanged since the 1930s. Contrary to the open-air museums developed by the Ministry of Tourism and Sport of Kazakhstan at the sites of Balkash Lake and Burabai in Central Kazakhstan, the village of Shabanbai Bi is not specifically organised to welcome visitors. From the point of view of the home-stay providers and the Nomadic Travel Kazakhstan operator, there are no special activities such as traditional cultural performances organised for visitors to portray Shabanbai Bi as a 'typical' Kazakhstani village. The Shabanbai Bi and Kanshengel villages, visited on the Kyzylarai and Tulip tours, respectively, are not considered by NGO coordinators and tourism operators as 'folk' villages, although authenticity is perceived differently by the home-stay providers in them. While the tourism operators see Shabanbai Bi village as objectively authentic, Kanshengel village is perceived as being staged for the sake of tourism. Home-stay providers emphasise that the visitors' stay in the guest houses, where they share a 'nomadic lifestyle' and interact with the local population, and that this can potentially enhance the authenticity of the tourism experience.

Home-stays, offered in yurts next to the Kanshengel village during the Tulip tour and in traditional guest houses in the village of Shabanbai Bi during the Kyzylarai tour were interpreted in various ways by the stakeholders. As yurts are part of both the nomadic lifestyle and visitors' perceptions of traditional nomadic culture before arriving in the country, some home-stay providers mention the need to reconstruct traditional yurts as a means to augment the perceived authenticity of the visitors' experience. This authenticity is negotiated and socially constructed between the visitors and the tourism operators who offer a tourism experience that reflects the contemporary socio-cultural Kazakhstani reality, with the culture presented in the villages constantly being reinvented. The concept of 'pseudo-events' or superficial and manufactured tourism experiences, as described by Boorstin (1964), finds some relevance in the Shabanbai Bi village when some Kyzylarai home-stay providers suggest the possibility of accommodating tourists in reconstructed yurts, a form of accommodation that has normally disappeared from rural villages. In the Kyzylarai tour, visitors are welcome to spend their time with their hosts in the guest houses, but some home-stay providers emphasised the possibility of building reconstructed yurts, depending on the tourists' demands. Here, the constructivist approach that relates to the meaning visitors can give to the object or event being evaluated in terms of its authenticity becomes important in both tours.

In the case of Shabanbai Bi village, relationships are developed between the local population and visitors when the visitors are invited to share a meal in the villagers' houses. Cross-cultural understandings between tourists and home-stay providers are favoured during evening meals at the guest houses in the form of a 'family feeling', where the level of intimacy between visitors and the host community members is high. The way Kazakhstani home-stay providers have organised the daily sites visits and the guest houses chosen to welcome visitors tends to favour intimacy between hosts and guests, and such a degree of intimacy is required if visitors are to fully appreciate the complexities of Kazakhstani eco-cultural heritage. The type of comfort expected in the Tulip tour clearly highlights antithetic aspects associated with the authenticity of the tourism experience: How to experience a traditional nomadic lifestyle without changing its meaning, for example by augmenting the level of comfort, is an interesting aspect. As one Kyzylarai tour home-stay provider mentioned, 'Yurts with solar panels, TV and fridges are part of local people everyday life, but it also allows us to cater [for] the needs of our visitors, especially more comfort for older tourists.' However, tourism providers do not see a certain level of customising the experience to meet tourists' demands for comfort as something that would lower their guests' perceptions of authenticity.

Performative spaces: nomadic food

All tourism providers interviewed mentioned that culinary traditions remain 'intact' in rural areas and the traditional table filled with dishes (*dastarkhan)* is perceived as objectively authentic. In Shabanbai Bi village, food traditions and recipes using horse meat (*bes barmak*, *kuyrdak*) and horse milk (*kymiz*) are acknowledged by home-stay providers to to have been handed down through the generations. The national dishes *bes barmak* and *kymiz* are served for visitors without any changes to the recipes during the Kyzylarai tour. One of the home-stay providers in the village of Shabanbai Bi observed:

> Our cuisine is very rich and national traditional food is the best expression of our culture; especially the way we are preparing horse meat and dairy products derived from horse milk. Moreover, the food we are serving for the visitors is the same as what we are eating ourselves.

Conversely, the yurt-camp organised in the steppes landscapes in South Kazakhstan for the Tulip tour is adjusted (food, levels of comfort in the yurts and proper toilets) especially for the needs of the visitors. The operators realise, however, that these adjustments may not entirely satisfy visitors:

> For this tour at first we have tried to meet the needs of Western tourists, by offering them popular Western dishes. But this year we slightly changed the direction and moved towards more traditional cuisine. And this is the first time we are cooking food outdoors.

In both cases, and despite a Westernisation of local food prepared by the organisers of the Tulip tour, the tourism operators clarified that the authentic part of the tourist experience can be found in *sharing* traditional meals prepared by the host populations in the villages.

The traditional *dastarkhan* prepared by the home-stay providers allows visitors to discover the conventional ways of cooking within the village and experience an 'authentic meal' with the local population. Thus, the impression of a genuine tourism experience is given by home-stay providers through culinary aspects of their cultural heritage, in particular when the visitors are given the possibility to participate in the preparation of the meals and have the recipes and experiences explained by the tourism operator or guides of the tours. The opportunity to be part of the cooking experience is provided spontaneously by home-stay providers as part of their daily lives in the villages. As tourists involved in active participation rather than observation are more likely to experience a sense of existential authenticity (Kim & Jamal, 2007), the intimate experience of sharing local culinary knowledge in villages or in yurts is perceived as an 'authentic encounter' by the majority of the home-stay providers, who explained that they do not need to stage the cooking of the meals specifically for visitors.

Discussion

Using a qualitative case study research approach, the study details that three themes authenticated by tourism providers can contribute to an authentic eco-cultural tourism experience for visitors: the geographical imagination (nomadic cultural landscapes), crafts purchased by tourists, and performative spaces (nomadic home-stays and nomadic food). The research shows that almost all the stakeholders involved in the development of Kazakhstani eco-cultural tourism emphasise the importance of ecological and cultural aspects in the visitor experience. For tourism providers, cultural landscapes and culinary traditions often constitute the main aspect of authenticity of the tourism experience being offered to visitors. While a majority of the home-stay providers emphasise the importance of nature and culture preservation, NGO coordinators highlight the notion of authenticity as being a unique and important feature associated with eco-cultural tourism practices and tours offered in the country. For government officials, a rebirth of nomadic traditions is necessary for tourism development purposes as it contributes to validate the visitors' romantic views of the nomadic culture and the Silk Road. Suppliers of tourism experiences in the Tulip tour conceptualise and shape the tourism experiences made available to visitors by supporting the commodification of some aspects of the Kazakhstani nomadic culture. In the Kyzylarai tour, some home-stay providers mentioned the need to reconstruct traditional elements of the nomadic culture shelters as a way to augment the perceived authenticity of the visitors' experiences. The organisation of staged yurt villages in the steppes is paramount to creating the 'feeling of being a nomad', a romantic view and representation by the visitors of the traditional nomadic culture. Despite a Westernisation of local food prepared by the organisers of the Tulip tour, the tourism operators clarified that the authentic part of the tourist experience can be found in the sharing of traditional meals prepared by the host populations in the villages. Authenticity is thus negotiated and socially constructed between the tourism operators and home-stay providers who offer a tourism experience that reflects the contemporary socio-cultural Kazakhstani 'reality'. While a majority of home-stay providers perceive cultural landscapes and culinary traditions as being objectively authentic, they perceive contemporary nomadic craft-making as a way of reviving the traditional Kazakh culture when the crafts are bought by the tourists.

A key managerial contribution of this research is the role authenticity plays in the planning and development of Kazakhstani tourism and local community participation and subsequent sense of empowerment. Wallace and Russell (2004, p. 236) argue that 'eco-cultural tourism acts as a model for how cultural and eco-tourism could be employed by local people to build an empowered, sustainable future in similar settings.' For Byrd (2007, p. 7) 'sustainable tourism cannot be achieved if imposed without regarding the stakeholders' interests'. Aas, Ladkin, and Fletcher (2005) and Jamal and Getz (1995) emphasised the importance of collaboration with stakeholders in shaping tourism development, implying that consideration should be given to each stakeholder group without one being given priority over others. According to Xie (2011, p. 185), allowing local communities to present their own culture themselves, without intermediaries being involved, 'can lead to a better tourism experience and create additional employment opportunities for minority people.' As Kazakhstani cultural landscapes have a strong significance to local populations, one of the ways to raise awareness of the cultural values of such sites is by empowering the local populations to provide stories about the places while tourists are in their guest houses. Explaining the cultural meanings behind traditional dishes or traditional artefacts, home-stay providers can communicate local stories that are perceived as 'genuinely' authentic by visitors. The Tulip and Kyzylarai tours offer different levels of commodification of cultural heritage and, more importantly, different tourism product offerings that reflect the organisers' different views on eco-cultural tourism development. While home-stay providers on the Kyzylarai tour are given more freedom and empowerment by the tour organisers to interact with tourists in Shabanbai Bi village, the tourism experience in Kanshengel village is more controlled by the Tulip tour organiser who offers a commodified version of the nomadic culture that limits contact and discussion between hosts and guests. Principles of tourism reciprocity imply that local communities will share traditional aspects of nomadic culture with the visitors when the hosts see benefits coming from tourism development. However, as Prideaux and Timothy (2008, p. 11) maintain, 'The power of tourism operators to select which attractions are patronised can sometimes result in a power transfer from the local community to commercial interests.' While the Nomadic Travel Kazakhstan operator allows the home-stay providers in Shabanbai Bi village to interact and organise the catering freely with their guests, Tulip home-stay providers are guided by the tour organiser, who frames the tourists' experience in the Kanshengel village. The structure of this second tour means that it is the tourism operator who determines how the families present authentic aspects of their cultural heritage, rather than the decisions coming from the local communities themselves. This has resulted in some of the 'authentic' aspects being staged to meet the desires of the visitors; for example, the re-building of yurts with higher levels of comfort next to the home-stays in Shabanbai Bi village. By explaining to home-stay providers what the key contributing elements of a visitor's authentic experience are, tourism organisers, in collaboration with government officials and NGO coordinators, can frame the boundaries of the tourism products and experiences made available to the visitors in a way that they feel is a genuine Kazakhstani cultural heritage. By helping to foster the network of home-stays in rural villages and identify the souvenir production of traditional handicrafts, local NGOs participate in the authentication and design of eco-cultural tours that can meet visitors' expectations of authenticity when travelling in the country. The governmental Kazakhstan Tourism Association (KTA) who already holds seminars for local communities to acquaint them with visitors' expectations has thus a determinant role to play in understanding the local populations' views and the development of future tourism products that take into account the visitors' expectations of nomadic traditions. A higher involvement by the local communities as to how they engage with visitors and in the eco-cultural tourism development of their villages would meet visitors' demand for authentic tourism experiences even if, as one home-stay provider said, 'It implies rebuilding yurts in the villages.' The organisation of staged yurt villages in the steppes is paramount to creating the 'feeling of being a nomad', a romantic view and representation by the visitors of the traditional nomadic culture.

The various levels at which the commodification of nomadic culture for tourism purposes is decided varies according to the stakeholders involved in the process. The local authorities appear to have essentialist conceptions of Kazakhstani cultural heritage, stating that an authentic tourism experience implies reifying and staging architectural and traditional elements of nomadic culture traditions as it was before the arrival of the Soviet people in the early 1930s. The extent to which tradition is negotiable and subject to often politically motivated invention is an issue that affects the politics of authentication of eco-cultural sites in Kazakhstani rural areas. Kazakhstani governmental policies for cultural evolution should focus on the revival of the nomadic culture's uniqueness and distinctiveness but without ignoring the impact of assimilation and acculturation policies inherent in the modernisation of traditional lifestyles and traditions. Negotiated identities and cultural hybridity involving the mix of modern and traditional aspects of the nomadic culture are also favoured by Kazakhstani officials and tourism providers, who recognise the opportunities that the revival of nomadic culture can create for the development of eco-cultural tourism in the country. The emergence and increasing development of home-stay guest houses in the Shabanbai Bi village is one of the possible models of development for Kazakhstani eco-cultural tourism that still preserves the authenticity of nomadic traditions. The commodification of Kazakhstani nomadic culture can be seen as a way to diversify the economy from a self-sustained nomadic culture to a more professionally trained tourism culture, a development that is necessary if the country is to welcome more culturally aware visitors.

Disclosure statement

No potential conflict of interest was reported by the authors.

References

Aas, C., Ladkin, A., & Fletcher, J. (2005). Stakeholder collaboration and heritage management. *Annals of Tourism Research*, *32*(1), 28–48. doi:10.1016/j.annals.2004.04.005.

Bhabha, H. (2004). *The location of culture*. London: Routledge.

Boorstin, D. (1964). *The image: A guide to pseudo-events in America*. New York: Harper and Row.

Buckley, R., Ollenburg, C., & Zhong, L. (2008). Cultural landscape in Mongolian tourism. *Annals of Tourism Research*, *35*(1), 47–61. doi:10.1016/j.annals.2007.06.007.

Byrd, E. T. (2007). Stakeholders in sustainable tourism development and their roles: Applying stakeholder theory to sustainable tourism development. *Tourism Review*, *62*(2), 6–13. doi:10.1108/16605370780000309.

Chambers, D. (2007). Interrogating the 'critical' in critical approaches to tourism research. In I. Ateljevic, A. Pritchard, & N. Morgan (Eds.), *The critical turn in tourism studies: Innovative research methodologies* (pp. 105–120). Amsterdam: Elsevier.

Cohen, E. (1988). Authenticity and commoditization in tourism. *Annals of Tourism Research, 15*(3), 371–386. doi:10.1016/0160-7383(88)90028-X.

Cohen, E., & Cohen, S. A. (2012). Authentication: Hot and cool. *Annals of Tourism Research, 39*(3), 1295–1314. doi:10.1016/j.annals.2012.03.004.

Creswell, J. W. (1998). *Qualitative inquiry and design: Choosing among five approaches*. London: Sage Publications.

Creswell, J. W. (2007). *Qualitative inquiry and research design: Choosing among five approaches* (2nd ed.). Thousand Oaks, CA: Sage Publications.

Denzin, N. K., & Lincoln, Y. S. (2000). *Handbook of qualitative research* (2nd ed.). Thousand Oaks, CA: Sage Publications.

Donaldson, T., & Preston, L. E. (1995). The stakeholder theory of the corporation: Concepts, evidence, and implications. *Academy of Management Review, 20*(1), 65–91.

Freeman, R. E. (1984). *Strategic management: A stakeholder approach*. Boston: Pitman.

Guba, E. G., & Lincoln, Y. S. (1994). Competing paradigms in qualitative research. In N. K. Denzin, & Y. S. Lincoln (Eds.), *Handbook of qualitative research* (pp. 163–194). Thousand Oaks, CA: Sage.

Hall, C. M., Williams, A., & Lew, A. (2004). Tourism: Conceptualizations, institutions, and issues. In A. Lew & M. Lew (Eds.), *A companion to tourism* (pp. 3–21). Malden, MA: Blackwell Publishing.

Hollinshead, K. (2006). The shift to constructivism in social inquiry: Some pointers for tourism studies. *Tourism Recreation Research, 31*(2), 43–58.

Hughes, G. (1995). Authenticity in tourism. *Annals of Tourism Research, 22*(4), 781–803.

Jamal, T., & Getz, D. (1995). Collaboration theory and community tourism planning. *Annals of Tourism Research, 22*(1), 186–204. doi:10.1016/0160-7383(94)00067-3.

Jamal, T., & Hill, S. (2004). Developing a framework for indicators of authenticity: The place and space of cultural and heritage tourism. *Asia Pacific Journal of Tourism Research, 9*(4), 353–372. doi:10.1080/1094166042000311246.

Kim, H., & Jamal, T. (2007). Touristic quest for existential authenticity. *Annals of Tourism Research, 34*(1), 181–201. doi:10.1016/j.annals.2006.07.009.

Kirshenblatt-Gimblett, B. (1998). *Destination culture: Tourism, museums and heritage*. Berkeley, CA: University of California Press.

Kolas, A. (2008). *Tourism and Tibetan culture in transition: A place called Shangrila*. Abingdon: Routledge.

Littrell, M. A., Anderson, L. F., & Brown, P. J. (1993). What makes a craft souvenir authentic? *Annals of Tourism Research, 20*, 197–215. doi:10.1016/0160-7383(93)90118-M.

MacLeod, N. (2006). Cultural tourism: Aspects of authenticity and commodification. In M. Robinson & M. Smith (Eds.), *Cultural tourism in a changing world: Politics, participation and (re)presentation* (pp. 177–179). London: Channel View Publications.

Neuman, W. L. (2009). *Understanding research*. Boston: Pearson/Allyn and Bacon.

Pernecky, T., & Jamal, T. (2010). (Hermeneutic) phenomenology in tourism studies. *Annals of Tourism Research, 37*(4), 1055–1075. doi:10.1016/j.annals.2010.04.002.

Phillimore, J., & Goodson, L. (2004). Progress in qualitative research in tourism: Epistemology, ontology and methodology. In J. Phillimore & L. Goodson (Eds.), *Qualitative research in tourism: Ontologies, epistemologies and methodologies* (pp. 3–29). London: Routledge.

Prideaux, B., & Timothy, D. J. (2008). Themes in cultural and heritage tourism in the Asia Pacific region. In B. Prideaux, D. J. Timothy, & K. Chon (Eds.), *Cultural and heritage tourism in Asia and the Pacific* (pp. 1–14). New York: Routledge.

Reisinger, Y., & Steiner, C. J. (2006). Reconceptualizing object authenticity. *Annals of Tourism Research, 33*(1), 65–86. doi:10.1016/j.annals.2005.04.003.

Rickly-Boyd, J. M. (2012). Authenticity and aura: A Benjaminian approach to tourism. *Annals of Tourism Research, 39*(1), 269–289. doi:10.1016/j.annals.2011.05.003.

Suntikul, W., Butler, R., & Airey, D. (2010). Vietnam's heritage attractions in transition. In V. T. Hitchcock, M. King, & Parnwell (Eds.), *Heritage tourism in Southeast Asia* (pp. 202–220). Copenhagen: NIAS Press.

Wallace, G., & Russell, A. (2004). Eco-cultural tourism as a means for the sustainable development of culturally marginal and environmentally sensitive regions. *Tourist Studies*, *4*(3), 235–254. doi:10.1177/1468797604057326.

Wang, N. (1999). Rethinking authenticity in tourism experience. *Annals of Tourism Research*, *26*(2), 349–370. doi:10.1016/S0160-7383(98)00103-0.

Werner, C. (2003). The new silk road: Mediators and tourism development in Central Asia. *Ethnology*, *42*(2), 141–159.

Xie, P. F. (2001). *Authenticating cultural tourism: Folk villages in Hainan, China* (Doctoral thesis). University of Waterloo, Waterloo, Ontario. Retrieved from https://uwspace.uwaterloo.ca/handle/10012/975

Xie, P. F. (2011). *Authenticating ethnic tourism* (Vol. 26). Bristol: Channel View Books.

Xie, P. F., & Wall, G. (2002). Visitors' perceptions of authenticity at cultural attractions in Hainan, China. *International Journal of Tourism Research*, *4*(5), 353–366. doi:10.1002/jtr.385.

Yang, L., & Wall, G. (2009). Authenticity in ethnic tourism: Domestic tourists' perspectives. *Current Issues in Tourism*, *12*(3), 235–254. doi:10.1080/13683500802406880.

Yin, R. K. (2009). *Case study research: Design and methods* (4th ed.). Thousand Oakes, CA: Sage Publications.

Governance and local participation in ecotourism: community-level ecotourism stakeholders in Chiang Rai province, Thailand

Nicola J. Palmer and Nipon Chuamuangphan

ABSTRACT

This paper takes as its starting point the assertion that community-based ecotourism is an activity in which not all members of a community are able to be involved (through barriers or factors of exclusion) and/or wish to be involved (through personal choice). This has implications for discussions about community-level stakeholders, governance and tourism development. Firstly, the paper explores the actors' interactions as they relate to the use of resources and environment. Within this theme are seen more specific sets of social relations and sustainability elements. These elements highlight key factors, beginning with the actors, the resource uses, and continuing with the influence of different resource management regimes, of power and authority, of networks of social relations, of patterns of governance, and of internal and external relations that occur with actors both inside and outside the local villages. The geographical context is three villages in the northern Thai province of Chiang Rai, a key international ecotourism area. The study is underpinned by use of a range of qualitative methods and considers the views of 70 key informants. Participation in tourism by Thai villagers is argued to reflect both ability to be involved (centred on land ownership and its restriction of livelihood opportunities) and active choice of involvement (the extent to which tourism is individually considered as a potential livelihood option). Involvement in community based ecotourism in a traditional hierarchical society like Thailand appears to be linked to an individual's social standing or general position in society and to be far from a matter of free choice. Governance – in the sense of social order, social coordination, social practices – is identified as having a key influence upon the ways in which participation in tourism occurs at a local level. Social status, legitimacy and power are highlighted as issues for further research in relation to further understanding the dynamics of community-based tourism development.

Introduction

There are many academic and practitioner studies that have examined community-based ecotourism, both in the broader context of developing countries (e.g. see Cater, 2006; Fennell, 2008; Roe, Goodwin, & Ashley, 2004; Scheyvens, 1999, 2002; Tosun, 2000) and within the specific context of Thailand (e.g. see Choibamroong, Laws, Richins, Agrusa, & Scott, 2011; Laverack & Thangphet, 2007; Novelli & Tisch-Rottsteiner, 2012). However, there is a dearth of material that presents the views or responses of community members towards working in tourism, participating in decision-making and having a voice within the development process once a decision to participate in tourism has actively been taken. Community-based ecotourism is often assumed to be a saviour of poor rural economies in the developing world (and to a lesser extent, the developed world). Recognition is lacking of the fact that not every member of a community is able to be involved or wishes to be involved in community based ecotourism or tourism *per se*.

It has been claimed that one of the key reasons why ecotourism and community-based tourism (CBT) are developed relates to an empowerment goal – the development of tourism as a livelihood option for communities where development options are limited and/or marginalised groups are being targeted in a society (Roe et al., 2004; Scheyvens, 2002, 2012). The development of CBT and ecotourism is often driven by international development funding aimed at achieving rural economic diversification, with tourism being identified as an alternative livelihood activity. It is apparent that the employment of community-based ecotourism (CBET) as an economic and business development tool is often encouraged or advocated in traditional, non-westernised cultures.

However, CBT approaches are necessarily dependent on contextual social structures and any type of tourism development therefore, must be considered on how it fits into existing institutional frameworks (comprised of a set of formal organisational structures, rules and informal norms). This suggests that generic models or approaches to CBT cannot be applied to achieve empowerment or address inequalities in communities (such as those directed at specific genders or ethnic groups), namely because CBT will be operating within a climate where issues of empowerment or inequality are engendered or embedded within local contexts. The influence of social structures, we argue, should not be dismissed in that involvement in tourism may be recognised to entail choices that are, at least, to some extent governed or constrained by wider social forces and structural inequalities that affect capacity for participation (Palmer & Chuamuangphan, 2017).

The contribution of CBT or ecotourism to economic and business development therefore cannot be considered without reference to the community and societal structures in which these developments occur (Higgins-Desbiolles, 2006; Kowal, 2008; Tosun, 2000). In particular, the dominance, power and control of certain actors in economic activities and as business owners within a community need to be considered as do capacity issues around entry to the CBT and ecotourism sectors. At local levels, distinctions are often made between businesses from 'inside' or 'outside' of local communities. These divisions challenge stakeholder management approaches that conceptualise group communities as cohesive social entities (Dunham, Freeman, & Liedtka, 2006).

This paper is interested in developing a better understanding of the involvement in ecotourism by communities in relation to three themes: appropriate livelihoods and ecotourism; social and resource equity and ecotourism; and decision-making and ecotourism management. These themes are conceptualised to be related to debates surrounding CBT (and ecotourism as a community based tool), ecotourism as an economic and business development tool, and cultural change and sustainability.

There is a need to understand the cultural contexts in which sustainability occurs. Resource management is affected by community and social values towards resources. Debates around sustainability are relevant since ecotourism is often used as a sustainable tourism tool. Yet it cannot be assumed that members of communities in lesser developed nations will hold similar notions of fairness to members of communities in more developed countries and, instead, these notions should be seen to be complex consequences of ideological norms often developed and sustained by particular political regimes. In traditional societies, which are often hierarchical and heavily role-based, then it is easy for 'outsiders' to highlight inequalities that appear to frustrate the principles of distributive justice. In tourism, this has often happened, particularly in relation to poverty alleviation and Pro-poor Tourism (Carbone, 2005; Hummel & van der Duim, 2012). However, there is a lack of research that considers non-western constructs of justice (in the broader context of perceived 'fairness') in relation to resource distribution in tourism at a community level of analysis (Dorjsuren & Palmer, 2018). This aspect is important in relation to our broader intention (through this paper) of contributing to better understanding of the involvement in ecotourism by community-level stakeholders.

Community involvement in tourism decision-making

There are many studies of tourism development that advocate increased community involvement and participatory planning, seeking to widen inclusion in tourism decision-making. Blackstock (2005) argues that structural inequalities within communities influence local decision-making and that trying to redress low and exclusive levels of involvement in tourism decision-making requires more radical intervention. The issue of *appropriate* participation in policy decision-making is complex and culturally-specific norms and societal values need to be taken into account. Only by gaining the perspectives of actors based within specific geographical case study areas can these issues be explored. It has been identified in a number of tourism studies that the existence of power elites and connections to the most powerful can affect the ability of local people to influence tourism decision-making. Blackstock (2005) shares Cheong and Miller's (2000, p. 381) observations of the relevance of wider structural inequalities in her observation that '*having the least control can translate into having the least involvement*'.

Conceptual framework

Prior to data collection, a conceptual framework was developed, influenced by three key theoretical approaches: ecotourism planning and management; sustainable development; and political ecology. This theoretical underpinning was related to relevant topics (social relations, ideologies of sustainability and ecotourism planning and management) and also other sub-topics in the study's conceptual framework (Figure 1).

Three key concerns or topics relevant to ecotourism planning and management are highlighted within the framework: the patterns of social relations relevant to sustainable development; the values and discourses of different actors about sustainable development, and the policies and practices of ecotourism planning and management that may be relevant to sustainability. It should be emphasised that the framework is not intended to be a rigid, pre-determined analytical framework, but rather it is intended as a broad, and fairly

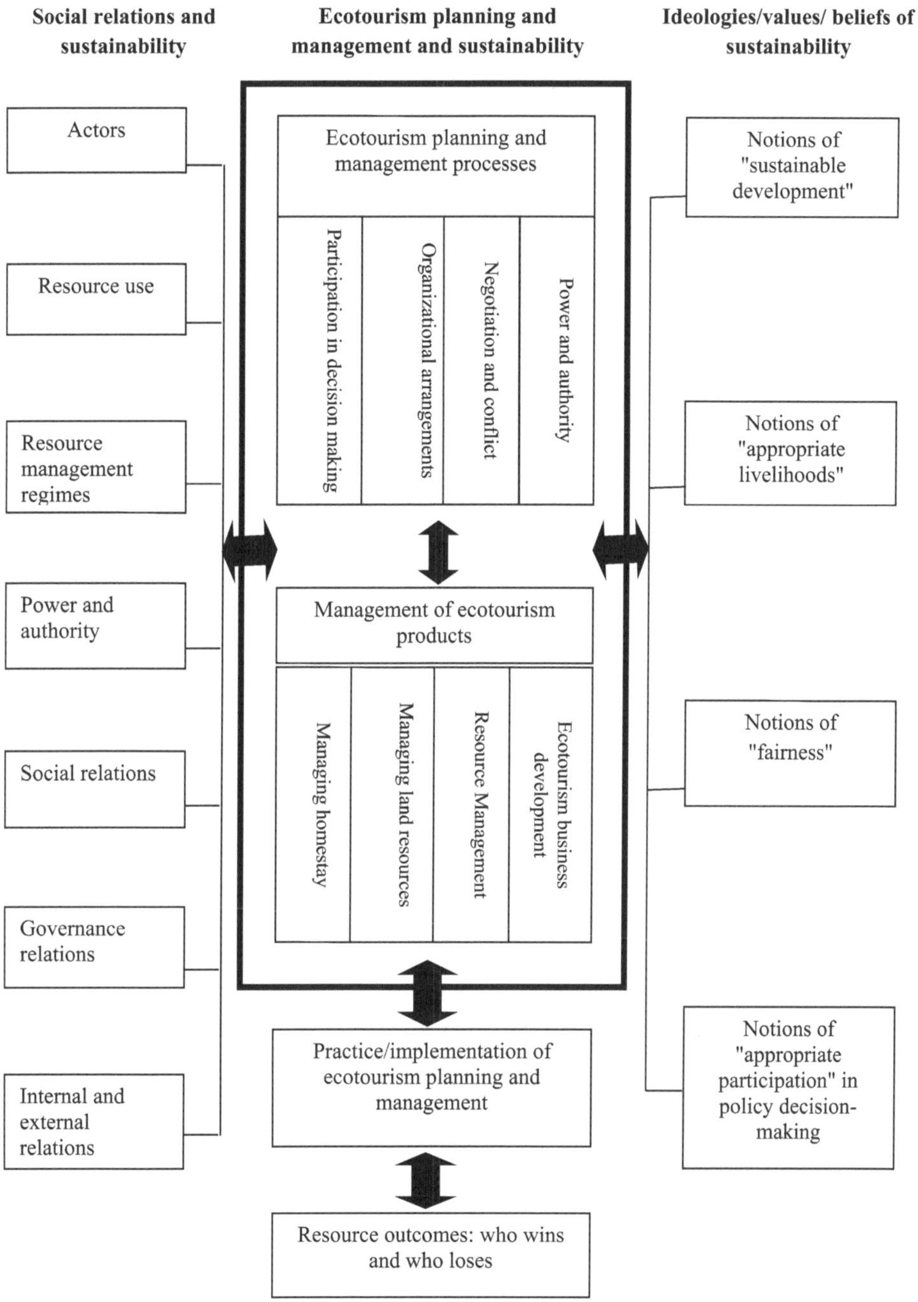

Figure 1. Conceptual framework to evaluate ecotourism planning and management and sustainable development.

loose integrative framework that links together social processes and environmental issues that affect sustainability.

Figure 1 shows that the first topics in the framework is social relations and sustainability, which explores the actors' interactions as they relate to the use of resources and environment, and to the local ecotourism products. Within this topic are seen more specific sets of social relations and sustainability elements. These elements highlight key factors, beginning with the actors, the resource uses, and continuing with the influence of different resource management regimes, of power and authority, of networks of social relations, of patterns of governance, and of internal and external relations that occur with actors both inside and outside the local villages. It is these factors that form the focus of this research paper.

The arrows in Figure 1 highlight the ongoing dynamic and dialectical interplay between the three topics; it can be said that the interactions between social relations and sustainability, the ideologies, values or beliefs about sustainability and ecotourism planning and management are manifested in each other and are closely inter-connected. Ecotourism, as a tool for sustainable development in developing countries, requires examination of organisational issues concerning control of the development process. The practical issues of ecotourism planning and management and sustainability are intimately connected with the themes of social relations and sustainability, and of the ideologies of sustainability. Social relations and the ideologies of sustainability profoundly affect the development of ecotourism planning and management, and these feed back into those relations and values. Local ecotourism planning and management may be directly influenced by what are considered appropriate processes according to local community perspectives. It may also be affected by relations of influence and power in the community. Therefore, it is important to focus on issues such as power and authority, negotiation and conflict, organisational arrangements, and participation in decision-making. This also facilitates consideration of who is able to wield the most power between the public sector, the private sector and the NGO sector.

The practice or implementation of ecotourism planning and management requires consideration of how strategies are converted into action around the development of land and other resources, around tourism business development, and around the organisation of homestays at a local level. Linked to this, there is the issue of local level strategies around employment and the sharing of benefits. These local level strategies must accommodate, or at least acknowledge, the local ideologies of sustainability as well as local social relations and sustainability. Ecotourism planning and management and sustainability needs to incorporate actor views on sustainability if ecotourism is to be able to truly promote sustainable development, as is often assumed by the governments of developing countries when ecotourism is selected as a development option. Thus the relevance of adopting an ethnomethodology was acknowledged in line with the study aim of examining actors' social relations and their responses to the role of ecotourism development within their local social context. Ethnomethodology, as an overall approach, encourages us to look at people's everyday ways of producing orderly social interaction. Using this constructivist strategy, the study places emphasis on human activities and experiences as based in the social relations, values and discourses surrounding ecotourism and sustainability.

Within the topic of social relations and sustainable development, there are three broad sub-topics. The first involves resource use, resource management, and the deliberations undertaken within communities concerning resources. The resources refer both to resources that support local ways of life and also to resources that more specifically support ecotourism in the area, such as community forests, local ways of life and indigenous culture, wildlife and national parks, and protected forest areas. The second sub-topic considers power and governance. It examines matters relating to the balance in power relations for communities, including issues around the power of local political leaders, as well as power in relation to external relations and actors. It also considers issues around socio-economic and political influence and control in the villages. The third sub-topic examines internal and external relations, notably the connections between village actors and the activities and actors involved from outside. This includes external investment in tourist facilities, the role of external tour operators, and internal and external involvement in the processes of ecotourism management and sustainability.

The framework presented in Figure 1 worked as a guide for the wider research study on which this paper is based, and it influenced the research methodology, and data analysis. It was intended to assist understanding and study of ecotourism planning and management and sustainability in a holistic, integrative manner which relates ecotourism to its socio-economic, political and environmental relations. It should be noted that the conceptual framework played a significant role in the design of the research methodology used in this study, it influencing the selection of research tools, the specification of the interview questions and the data interpretation.

Research case study and method

A case study approach allows for consideration of agency and structure and choices and constraints within the distinctive situation, of the complex multivariate conditions found in those situations, and of the multiple relevant sources of evidence (Yin, 2003). This study was intended to focus on the structural dimensions of the social relations within local communities, and between those communities and actors outside, in relation to ecotourism planning and management, and it also considered what the actors felt about those relations and their way of life. Thus, the value of examining specific geographical and community cases was noted.

Rural areas of Thailand are noted for the remarkable growth and significance of tourism, in general (Choibamroong et al., 2011), and ecotourism, in particular. As a geographical research case, the province of Chiang Rai in northern Thailand is a key international ecotourism area with a range of tourism development stages and community characteristics to examine. Three rural areas of Chiang Rai were used as case studies: Rong Born village; Yang Kham Nu village; and Ruammit village (see Figure 2).

The three case study villages are distinctive in that their communities are made up of indigenous hill-tribes, but they share a reliance on agriculture as the principal economic sector and livelihood activity. The patterns and processes of ecotourism planning and management in each of the three rural areas were examined prior to the fieldwork phases of this study with a strong focus on these differing local contexts.

Differing management regimes were in place: Rong Born involved territorialised state control through its national park status and through the powerful state forest service; Yang Kham Nu was characterised by greater local community influence and control; and Ruammit was characterised by both local and external tourism companies and by increasing local government involvement (as part of a Thailand-wide attempt to strengthen local government). All three of the villages shared the same structure of local governance. Village leaders were democratically elected by local communities and there were no gender and ethnic restrictions on becoming a leader. Financial management of the villages

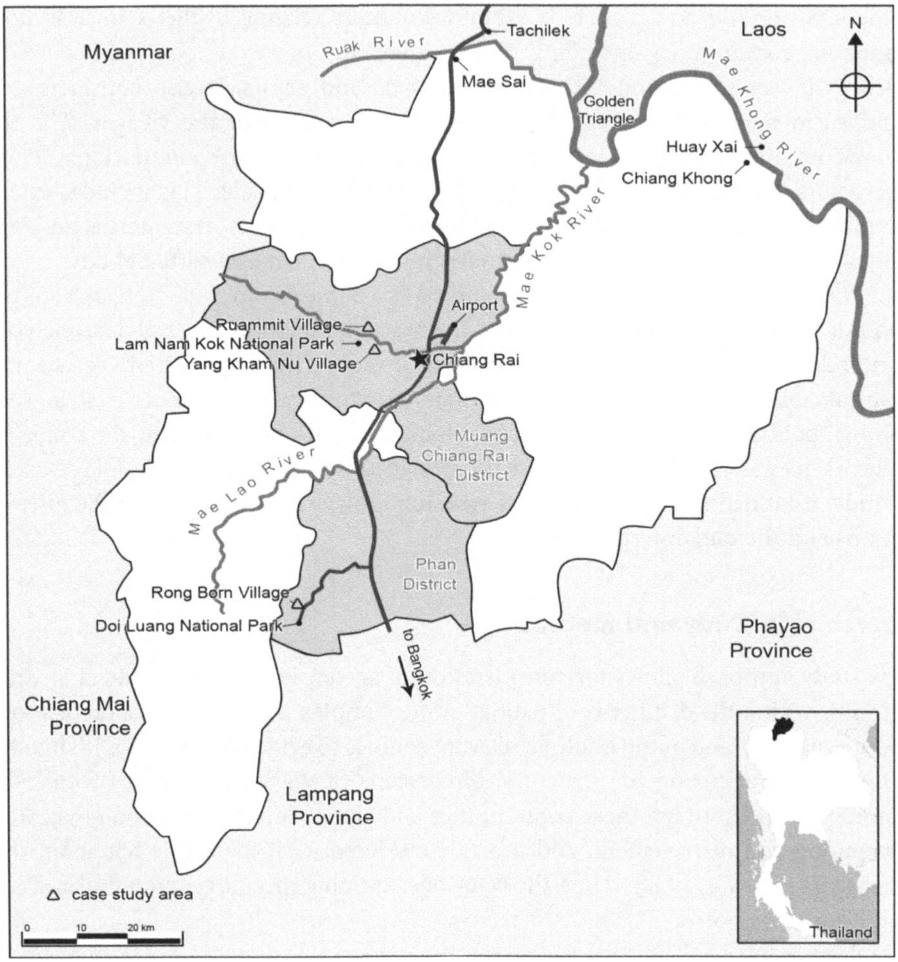

Figure 2. Chiang Rai Province, Thailand and the location of the three case study villages. Source: The authors.

was undertaken by these village leaders who played an administrative role. Village meetings, the official forum for village leaders to consult with their communities were found to vary in frequency – seldom held in Rong Born but held once a month in Yang Kham Nu and Ruammit.

There were also differences in terms of ecotourism development trajectories in the three villages. The most established tourist destination area was Ruammit village where trekking tours and elephant trips had been operating since 1975. Tourism activity in the local economy had been largely controlled by external travel agencies who brought tourists into the village and consequently, took a significant portion of the economic returns. However, a wide range of tourist services were also provided by villagers, in cooperation with travel agencies and also the Tourism Authority of Thailand (Chiang Rai office). In contrast to the other two villages, most tourists visiting Ruammit village were international tourists. In Rong Born, the local forest had been destroyed through over-use in the past and, with the intervention of the village leader and advice from an NGO and village elders, a community forest had been developed in 1992 which formed the basis of the village's ecotourism offer since 2001. Ecotourism was launched in Yang Kham Nu village at the beginning of the 1990s. At provincial level government Doi Hang Tambon Administration Organisation (TAO) planned and developed tourism, focusing on facilities and infrastructure for tourists, for example, developing road access to the sites and providing accommodation for visitors (p.c. Doi Hang TAO, 2007). The TAO directly received income from tourists and paid some local people to become staff at service places. In this village the TAO also promoted homestay accommodation.

Three intensive periods of fieldwork were undertaken in the villages spanning three consecutive years by Chuamuangphan, a Thai native speaker. A range of qualitative data collection methods were employed, including observation and non-participatory attendance of village meetings, 70 personal face-to-face semi-structured interviews and document analysis. Of the 70 key informants interviewed 50 represented local community members: 12 from Rong Born; 12 from Yang Kham Nu; and 26 from Ruammit. The remaining 20 represented national, provincial and district level actors, non-government officials, tour operators and representatives from non-governmental organisations involved with ecotourism planning and management. The 50 community members included village leaders, farmers, elderly and young people, tourism business owners, guides, and accommodation service providers (homestay service), religious leaders and housewives. Collectively, they may be recognised to all be representatives of 'community-level tourism stakeholders'. A snowball sampling technique (Denscombe, 2003) was used in their recruitment.

Figure 1 served as a tool to guide the themes for the interview schedules used in the study fieldwork. It provided a means of ensuring that the semi-structured interviews focused on issues that were relevant to the wider study aims and objectives. The framework also aided comparability between the actor interviews, which were adapted according to actor education levels and their degree of familiarity with conceptual language, to ensure that similar issues were examined by the different respondents. The data was collected by Chuamuangphan, a native Thai speaker, and was translated prior to transcription. Translation from Thai to English took place during the transcription process. This enabled Palmer to participate in the process of confirmability in line with the pursuit of 'trustworthy' qualitative research (Guba, 1981; Shenton, 2004). The collected data was analysed using content analysis, a technique that provides new insights and increases researcher understanding of particular phenomena. This was identified to be especially relevant here in relation to examining local responses. Thematic analysis was employed, where the coding scheme was based on categories designed to capture the dominant themes present in the text, (Franzosi, 2004, p. 550). This paper focuses on the discussion of three key themes: discourses around appropriate livelihoods and ecotourism; discourses around social and resource equity and ecotourism; and discourses around decision-making and ecotourism management.

Discourses around appropriate livelihoods and ecotourism

Land ownership was a key factor affecting livelihood decisions. It was found, in all three villages, that land ownership was a major influence on perceptions of appropriate livelihoods, particularly in relation to involvement in ecotourism. The people who were landless were more likely to move into tourism because it provided them with a means of livelihood over which they had control. This observation suggests that involvement in tourism was utilised in the Thai villages as a means of securing control or a perceived route to empowerment. It supports the ideas of Blackstock (2005) that structural inequalities within communities influence local decision-making; in this case it may be argued that marginalisation and inequality drove certain community members into the tourism industry.

A tourist guide in Yang Kham Nu said,

> 'I have no land for farming. I work as a trekking tour guide ... I think I have enough money for my family ... such as I can pay the educational fee at a university in the city for my daughter. I am satisfied with this job.'

A youth at Yang Kham Nu observed,

> 'My father has got land for farming ... but we are a big family ... and I am happy to work in a hotel in the city ... Yes, I ride my motorcycle to work and return home because it is not that far from here [his home].'

Additionally, a shop owner at Ruammit stated that:

> '*We* [the interviewee and his sister] *had no land for farming at our hometown ... My sister moved to work here and opened her small souvenir shop in this village* [Ruammit] ... *We are happy to work here ... I can get higher education ... I am studying at a university in Chiang Rai*'.

Thus, participation in tourism as a livelihood appeared to be viewed as 'a means to an end', an economic livelihood chosen, at least in part, as a means of financing or supplementing a standard of living. In terms of livelihood options, farming was continually positioned against tourism as a way of making a living, perhaps unsurprising given the heavy economic dependence on farming in each of the three villages.

However, control over this factor was perceived to be limited and legally-constrained. Tribal people felt that it was very difficult to own their own land because they had migrated from place to place in the past, pre-settlement in villages such as Ruammit and Yang Kham Nu. Partly as a consequence, they regarded themselves as poorer than the indigenous Thai land-owning, people. The groups of tribal people at Ruammit, for example, said that they did not own land because, as tribal people, they lacked some of the rights of people who had Thai national identity. One of the village leaders explained some of the complexities of and implications of rules around land ownership:

> '*There were some business people from outside who came to the village to buy land from the villagers ... For example, my relative, she wanted to sell her land to the business man ... I did not agree with her ... I told her to compare the good and bad sides after selling the land and that she would become landless ... Moreover, some tribal people around the sub-district had no Thai nationality card ... so, it is difficult for them to buy land for themselves*'.

Lack of land ownership did not however prevent tourism involvement but, rather, affected the nature of that involvement, prompting entrepreneurship and self-employment activities. Landless people, especially some of the tribal groups in Ruammit, had become very involved in tourism businesses in the village, especially selling souvenirs to tourists. The indigenous Thai people acknowledged that tourism businesses brought economic benefits to the owners of those businesses often indirectly linked to the tourist trade.

Tribal groups, such as the *Akha* and *Hmong* in Ruammit, did not have their own land and mostly rented a small parcel of land for running their souvenir shops. However, some of these community members worked simultaneously as shop owners in Ruammit and as agricultural workers in their previous village of main residence. They reported that they would go back to work on their farms in the rainy season (from May to July yearly) and they returned to Ruammit after harvesting their crop to re-open the shop again. Thus, their status and identification as community members was complex and split between two villages. Here, the idea of communities as being cohesive social entities was clearly challenged in line with the sentiments of (Dunham et al., 2006).

In contrast to Ruammit, in Rong Born and Yang Kham Nu the villagers tended to prefer working on farms rather than getting involved in tourism. One influential factor here might be the level of tourism maturity in their villages – tourism was newer (less established) in their villages in comparison to Ruammit. This meant that villagers had less experience of and fewer ideas about tourism as a livelihood compared to agricultural work. One farmer in Rong Born, for example, stated a preference for farming as a more secure livelihood than tourism:

> *'Yes, I have my own farm ... I like to work on farms ... I do not think we can do any tourism business in the village because there are only some people who come to the community forest once or twice a month'.*

Assessment of tourism in comparison to other livelihood options on the basis of personal capabilities was also apparent. One of the farmers in Ruammit, for example, stated, *'I like to work on a farm rather than working in a tourism job ... I do not think I can do any tourism job ... I am too shy to be a seller ... It is hard for me to work in tourism'.*

Involvement and participation in tourism as a livelihood was found to be less linked to what tourism could offer them (Walpole & Goodwin, 2001) but was linked to national identity status and lack of resources to make alternative livelihood choices. Material circumstances and cultural values were identified as instrumental to the ability of local people to make alternative livelihood choices. In this context, the relationship between influence, control and participation was seen to be affected by capability and, in some cases, clearly reflected restrictions over opportunity to make other livelihood choices; the question arises as to the extent to which tourism in this context is a lesser-favoured or secondary livelihood option rather than an economic activity of choice?

Discourses around social and resource equity and ecotourism

Notions of fairness around socially-distributed benefits from ecotourism and exploitation of resources for ecotourism vis-à-vis other community activities were found to be inextricably linked. In Rong Born and Yang Kham Nu there was a strong focus on environmental conservation in day-to-day village activities, and this conservation focus had subsequently attracted tourists to see their conservation practices and community forest work. In Rong Born, commercial ecotourism development had led to environmental degradation, and as a result of this, conservation concerns had been given more prominence. Thus, tourism was recognised to act as a tool for enhancing the value of environmental resources and, by virtue, providing wider benefits to the community as a whole. It also acted as a driver for environmental protection. This was one reason why although not all villagers worked in tourism nor received direct economic benefits from tourism, resentment was not commonly expressed towards those villagers who did.

However, amongst those who did work in tourism the distribution of benefits or outcomes for them was questioned. A housewife at Rong Born argued that:

> *'I and many of housewives love to provide services for visitors, but we cannot work for free … We have jobs to do to gain income for my family … This time if the village leader wants a cook for visitors, she has to pay for housewives … she has now paid round 100 Baht a day … It is a good deal'*

In Rong Born, villagers had joined in with tourism activities in the village expecting to get a return from their participation. In fact, they got nothing and it involved an opportunity cost – they lost their benefits from farming or labouring at that time. One farmer at Rong Born bemoaned this situation:

> *'Think about yourself* [the interviewer]. *When you work almost a day in welcoming visitors activities … Take them to the trail [community forest] … But I got nothing … I have two children to get to school … I have to pay for them for transportation, lunch and for a snack … If I work for the village … what about my family? They have to eat and want money to support their life'.*

The external actors in this context refer to tourist organisers or tour operators based outside of the villages and they were involved more explicitly in Ruammit where tourism was more established. These actors had brought tourists to the village, they had certain transactional and thus, institutionalised relationships with the villages, and, at times, the relations between the external actors and the local community led tourism in a positive or negative direction. For example, individual commercial groups based outside the village led tourists to Ruammit for trekking, on half-day tours or for a few hours around the village:

> *'Tourists often came to the village via tour operators and they had just looked around the village without doing any shopping … They [tourists] just looked around the village for elephant riding, watching hand made products and seeing our [villagers] local ways … In the past, they [visitors] came for days, spent time with local people, stayed overnight with us … Yes, we got more money than now'* (Shop owner, Ruammit).

Thus, this change provided fewer benefits for local people when tourism was brought to the village by tourism organisers. A sense of injustice was expressed as exemplified by the view of one farmer at Ruammit:

> *'People who are involved in the tourism business should retain the benefits of their works … Like small local shops in the village, they have sold souvenirs or meals to tourists, then they should get the benefits [money] from the visitors'.*

This, of course, had implications for any expectations that tourism should be a community-based activity. There was tolerance for individualism, reflecting the significance of livelihood income at an individual household-level as well as at a community level.

When Yang Kham Nu provided a welcome event for tourists in the village based on a traditional indigenous Thai family-based tradition this provided only a limited range of benefits for community members and there was some local dissatisfaction with the exploitation of local 'private sphere' practices. One farmer at Yang Kham Nu stated:

> *'I feel unhappy with the show of the villagers for tourists, such as the tying of holy thread around the visitor's wrist. It is our way to highly respect our own family. However, we also want to satisfy our tourists and we want more of them to come to the village ... We will get more money from them'.*

The social status of tribal groups emerged as a fundamental, underlying issue. Indigenous Thai groups appeared to be more influential and powerful, having a higher social status than other ethnic groups. This is partly because indigenous people believed that tribal people were minority groups lacking Thai national identity. Moreover, status, in relation to resource equity and fairness around access to resources, also emerged as an issue. Some of the tribal groups were believed to have destroyed the forest for crops and logging purposes when they first arrived in the area in response to lacking farm land ownership and in an attempt to find livelihoods through logging. There was a small amount of cutting down of the forest in Ruammit and Rong Born by these ethnic groups, for example, before community forests were set up in their villages. This had created some tensions between villagers:

> *'Tribal people destroyed the forest* [pointed to a forest area] *... because they were landless and they were poor ... Like tribal people nearby our village, they had been moved out from a protected area and they asked to settle down on the side of Doi Luang National Park'* (Farmer, Rong Born)

> *'The indigenous people are richer than us ... they have good facilities ... perhaps, their ancestors had left them good land and things ... or they had higher education compared to us ... we are far away from the city'* (Housewife, Yang Kham Nu).

Resource management emerged as a key issue in relation to community views around social and resource equity and tourism. Politics and power inequities were evident in relation to access to natural resources, exploitation of resources for ecotourism purposes and subsequent distribution of benefits from resources in line with the findings of political ecology approaches to the study of tourism (Cole, 2012; Gossling, 2003; Stonich, 1998).

Discourses around decision-making and ecotourism management

The village leaders said that they welcomed all villagers to join in the tourism activities yet this was difficult because people who had never been concerned with tourism, such as farmers, did not want to be involved in tourism activities. Participation appeared to be affected by the exercising of choice, at least in terms of economic cost–benefit analysis in relation to livelihoods pursued by those owning land (the key to access to opportunity in the communities under study). In this respect, some level of power and control of community stakeholders to take part in ecotourism decision-making was evident. Opportunity to participate was observed to exist, however this appeared to be limited by community stakeholder understanding of the scope and value of their participation.

At the time of the fieldwork, the Thai government had been encouraging local people to become active participants in their own village planning under its '7th National Economic and Social Development Plan and Creative Economy (1992-1996)'. This encouraged communities to engage in managing their local resources, most notably community forests (p.c. National Economic and Social Development Board (NESDB), personal communication, 2008, Bangkok, Thailand). However, the government had not focused on what participative role local people would play in terms of local empowerment and local resource management for community development. In the three villages, the village leaders and the external government actors had sought ways to encourage greater local participation in village planning. An administrative member of the National Park Department in Bangkok said, '*... Yes, we are doing and encouraging them to join in ... But, sometimes the local people needed to be explained about the roles of local participation*'.

A key reason why they did not always know and understand their role in participation was the control exercised by their leaders and governors. Some local people felt that they had no opportunities to question or disagree with the planning for their villages regarding the decisions made by their local leaders. They did not feel like they had a voice in decision-making and a lack of trust was apparent. At Rong Born a housewife stated:

> '*Almost all of the projects within the community were planned by our leader and the leader team ... We (as a villager) sometime found it difficult to reject the projects ... Because, we did not want to have any conflict with our leaders and the projects were done for us ... no point to argue with*'.

Similarly, in Ruammit a shop owner remarked:

> '*The* [village] *leader is lazy. He just does and runs projects for the village for himself ... He has a restaurant there* [in the village] *and his sister is the elephant camp manager ... I have not seen him do a good thing to bring tourism to the village ... No, no I have never heard about tourism plans or there has not been a meeting about tourism plans or any project about tourism ... I cannot argue like this to him directly ... I do not want to get into conflict with them, and I am happy with this [situation]*'.

As a consequence, lack of community participation in tourism decision-making caused local people to be unclear about the ideas and concepts behind important developments for the villages and about the local benefits, such as the likely income from tourism for the village. In practice, plans for the villages were set up by leaders without participation from all villagers or involvement of all community members in local decision-making and tourism management. Village meetings were considered to be (and observed to be) focused on information dissemination rather than a means of facilitating consultation and active engagement. Community and societal structures were found to shape and constrain the contribution of community members to ecotourism, in turn, impacting on wider economic and business development decisions (Higgins-Desbiolles, 2006; Kowal, 2008; Tosun, 2000).

Participation in tourism decision-making was also influenced by relations between the villages and external government actors. Changes in government policy were interpreted with respect to how the villagers and their welfare were valued by those with higher levels of authority. For example, a new regulation imposed by the National Park banned local people from entering local forests to collect any wild products from the protected area. The introduction of this policy coincided with a change in National Park leadership. Local people construed that they were not allowed to collect the wild products just because the new National Park leader did not want them to go in. The Rong Born villagers spoke at length about how both their personal relations and the changes of policy affected their relations with the Doi Luang National Park, at the edge of their village. The current leader was perceived to be 'unhelpful' and 'unfriendly' to the villagers. Thus, the influence of leadership on participation was apparent.

Local people felt that they had good relations with local and national government when they clearly received benefits from the government. These benefits were assessed not just for individuals but for the support of village development as a whole. Villagers in Ruammit, for example, respected and considered their local government TAO (Mae Yao Tambon Administration Organisation) positively because the TAO supported them with a local tourist guide training project and this had enabled the forging of good relations between the village and the TAO. This was cited by interviewees as an example of where they felt that local involvement in tourism had been encouraged.

One of the research methods used in the study of the three villages was non-participatory observation and this included attendance of village meetings. These meetings were officially positioned as a means of enabling local community participation in ecotourism decision-making around ecotourism management. People attended the meetings as social occasions and also because they did not want to be penalised if they missed meetings three times in a year (as was local policy). Also, if they missed the meeting, they felt that they would not get any benefits from the village related to the projects or initiatives that were discussed in the meeting. The penalisation issue was explained by an elder at Ruammit:

> '*... Before joining the meeting. It is for all members to know what will be happening in the village and all news and information are useful for them to know ... Yes, the penalty has a good result to make members come for the meeting*'.

The long duration of the meetings (up to three and a half hours) meant that the villagers often felt that it was boring to stay to the end of the session. The locals often only wanted the very important issues to be discussed in the meetings, rather than having all issues covered. In fact the process often became one of just informing all the villagers about what was happening, rather than about making decisions.

Participation and involvement was often understood in relation to wanting to be involved to gain more benefits from the tourism promotion of the village. An example of this was provided by a shop owner at Ruammit:

> '*Of course, we want to be part of tourism involvement here* [at Ruammit] *... We have attended all the village meetings ... I have never missed the meetings ... I want to know about how tourism is going to be managed in a good way ... If there are more tourists, I could get more income from these tourists*'.

Relations that connected the villages and external tourism organisations, (such as TAT), were not as good as they should be. For example, many tourists and external tourism organisations had not known that Rong Born had opened its community forest conservation project for tourism, while Rong Born villagers had not known that TAT could help them to promote their community, citing TAT's lack of awareness and support for their annual 'Elephant Day' as an example. In Ruammit, the Mae Yao TAO had prepared both one-year and five-year plans to develop and improve ecotourism management in the village and rural areas around the village. All of the plans were about infrastructure and tourism facility improvement but they neglected to address sustainability in ecotourism management. In addition, the planning process did not have local participation. Rather than having discussions with local people, only leaders of teams from both the village and the local government were involved. The local villagers said tourism planning and management came from their leaders and the TAO. One shop owner at Rong Born said '*All plans for the village came from the leader team … I do not want to argue with them … the plans they have done are not bad … nobody wanted to argue with the leader's team*'. Thus, the plans were not felt to be representative of all local people's ideas but rather perceived to be imposed by virtue of local governance structures.

There have been major changes in the government of Thailand over many years, and this has led to a questioning of the authority of government officials. Despite this, the local villagers generally continued to respect people with status, including government officials. People praised the government as admirable and effective because it was the government that had provided them with 'One Million Baht for One Village' projects – low interest loan schemes intended to provide villages with a budget (£15,385 approximately) to enable local people to borrow money to improve their work. A religious leader at Ruammit commented:

> '*It is good if leaders or TAOs have good cooperation … in between villages and also in between a community … for example, Doi Hang TAO* [where Yang Kham Nu is) *and Mae Yao TAO* [where Ruammit is) *are planning to manage a new route for tourism between these TAOs … I think it must be a good programme … for visitors and for tourism*'.

The village leaders had a highly influential role and this risked negative repercussions for community involvement in ecotourism planning. The elected village leaders had considerable concentrated power, and they made their own decisions to select other people in the village to assist them in managing the village, creating a highly centralised form of paternal control over the village which was a way to keep administrative control over the village. For example, when they contacted the government, their leaders would act as intermediaries to help them negotiate with key people:

> '*I like having a short cut when contacting key people, especially government officers in the district. They carried a message from their leader and it helped to get faster service from the government officers … And I succeeded in what I intended to do … it is a short cut to be successful more easily*' (Shop owner, Rong Born)

In general, however, it was found that villagers often had limited understanding of their potential to participate in ecotourism decision-making.

There were mixed views around that extent to which villagers wished to be involved in tourism decision-making and who they felt should be most involved. Views were linked to ideas around notions of fairness in terms of social and resource equity in ecotourism. Local governance structures – namely, the role of the village leader and the effectiveness of village meetings as consultative mechanisms – appeared to be the main focus for structural inequalities (Blackstock, 2005; Leach, Mearns, & Scoones, 1999) within the three villages. In theory, the village meetings and NGO initiatives provided opportunities for local communities to gain involvement (Ashley & Roe, 1998) but barriers and constraints were revealed in terms of power and influence (Cheong & Miller, 2000).

Conclusions and recommendations

This paper provides community-level insights into local perspectives on ecotourism development in rural northern Thailand and this contextual contribution is important. The paper also holds relevance to three particular academic debates that extend beyond the specific geographical focus: the employment of ecotourism as an economic and business development tool; the importance of cultural change in sustainability; and appropriate levels of community involvement in tourism decision-making. It may be concluded that, in the three case study contexts, ecotourism was not a livelihood of choice but participation was governed by land ownership reflecting social status and citizenship, under legislative control. This challenges the ability of ecotourism to be developed as a community- based tool. It was believed that most benefits from tourism, mainly economic, should go directly to those working in tourism although the community as a whole should receive wider benefits in view of tourism's exploitation of community-owned resources, nature and culture in particular. This highlights 'legitimacy' as an important concept within the cultural context in which sustainability occurs. Involvement in tourism decision-making was recognised to be governed by local power structures and local politics and local community members saw little ways of influencing or working beyond locally-determined hierarchical processes. Political power as wielded by village leaders may be recognised as a threat to attempts to widen inclusion in tourism decision-making. A need to consider local responses to the concept of distributive justice and fairness in relation to resource distribution was supported - particularly in terms of legitimacy (Della Fave, 1980). Novelli and Tisch-Rottsteiner's (2012) suggestion that those working in tourism – mainly hill tribes – should be acknowledged and rewarded for their contributions was also corroborated.

Collectively, the findings indicate that local participation in tourism needs to be recognised to reflect wider social structures, values and practices. Governance – a requirement for knowledge; thought; the application of power, resources and rules; and also coordination and cooperation among numerous actors (Bramwell, 2011) – may be identified to have a key influence upon the ways in which participation in tourism occurs at a local level. Ecotourism can only operate as a mirror of society and community. It does not bring inherent benefits or dis-benefits and the (local) impacts of ecotourism should not be viewed in isolation from (local) social structures, values and practices.

One limitation of this paper is that the study draws on research evidence dating from the past. There is a need to consider updated research evidence and examine the extent to which social changes have occurred in terms of community structures, values and practices and tourism development. According to the UNCHR (2015) little has altered since the timing of the fieldwork and ownership of land in Thailand remains legally restricted to Thai citizens or Thai companies which have Thai nationality (Chaninat & Leeds, 2018; Cumminsky, 2010; UN Statistics Division, 2013). The 'Thailand Human Development Report, 2014' (UNDP, 2014: 81) draws attention to the observation that, in Thailand, '*Members of a village share a place*' and the impact of identities on community-level tourism stakeholders affects ways of seeing and ways of doing. These fundamentally underpin participation and involvement in ecotourism and add support for the adoption of interpretivist methodologies to enable community-level tourism stakeholders to represent themselves. 'Welcome to Chiangmai and Chiangrai', (recognised as a credible source by Time Magazine Asia), reports that local village life in northern Thailand is changing slowly with technological advancement but draws attention to the persistence of social structures, community values and practice as found in this study (Welcome to Chiangmai and Chiangrai, 2014).

In the context of traditional communities challenged by issues of resource scarcity such as the three villages in Thailand (and it may be argued other communities in lesser developed nations) there is also a need for researchers to further investigate the complex interdependencies between social structures, values and practices, resources and livelihoods. This paper acknowledges the value of political ecology and wider governance approaches to further examine social status, legitimacy and power in relation to ecotourism development and to assist academic, practitioner and policy-maker understanding of ecotourism as a community development tool and to progress stakeholder analyses.

Disclosure statement

No potential conflict of interest was reported by the author(s).

References

Ashley, C., & Roe, D. (1998). *Enhancing community involvement in wildlife tourism: Issues and challenges* (Vol. 11). London: IIED.

Blackstock, K. (2005). A critical look at community based tourism. *Community Development Journal, 40*(1), 39–49.

Bramwell, B. (2011). Governance, the state and sustainable tourism: A political economy approach. *Journal of Sustainable Tourism, 19*(4–5), 459–477.

Carbone, M. (2005). Sustainable tourism in developing countries: Poverty alleviation, participatory planning, and ethical issues. *The European Journal of Development Research, 17*(3), 559–565.

Cater, E. (2006). Ecotourism as a western construct. *Journal of Ecotourism, 5*(1-2), 23–39.

Chaninat & Leeds. (2018). *Buying land and property in Thailand.* Retrieved from http://www.thailand-lawyer.com/land_purchase.html

Cheong, S. M., & Miller, M. L. (2000). Power and tourism: A foucauldian observation. *Annals of Tourism Research, 27*(2), 371–390.

Choibamroong, T., Laws, E., Richins, H., Agrusa, J., & Scott, N. (2011). A stakeholder approach for sustainable community-based rural tourism development in Thailand. *Tourist Destination Governance: Practice, Theory and Issues*, 173–186.

Cole, S. (2012). A political ecology of water equity and tourism: A case study from Bali. *Annals of Tourism Research, 39*(2), 1221–1241.

Cumminsky, J. (2010). *Thai land, Thai owners.* Retrieved from http://www.property-report.com/thai-land-thai-owners/

Della Fave, L. R. (1980). The meek shall not inherit the earth: Self-evaluation and the legitimacy of stratification. *American Sociological Review, 45*(6), 955–971.

Denscombe, M. (2003). *The good research guide: For small-scale social research projects.* Milton Keynes: Open University Press.

Dorjsuren, A., & Palmer, N. J. (2018). Equity in tourism development: Procedural (in)justice and distributive justice in Mongolia, East Asia. *Asian Journal of Tourism Research, 3*(1), 60–90.

Dunham, L., Freeman, R. E., & Liedtka, J. (2006). Enhancing stakeholder practice: A particularized exploration of community. *Business Ethics Quarterly, 16*(01), 23–42.

Fennell, D. A. (2008). Ecotourism and the myth of indigenous stewardship. *Journal of Sustainable Tourism, 16*(2), 129–149.

Franzosi, R. (2004). *From words to numbers: Narrative, data, and social science* (Vol. 22). Cambridge: Cambridge University Press.

Gossling, S. (2003). *Tourism and development in tropical Islands: Political ecology perspectives.* Cheltenham, UK: Edward Elgar.

Guba, E. G. (1981, June). Criteria for assessing the trustworthiness of naturalistic inquiries. *ERIC/ECTJ Annual Review Paper, 29*(2), 75–91.

Higgins-Desbiolles, F. (2006). Reconciliation tourism: Healing divided societies. *IIPT Occasional Paper, 7*, 35–44.

Hummel, J., & van der Duim, R. (2012). Tourism and development at work: 15 years of tourism and poverty reduction within the SNV Netherlands development organisation. *Journal of Sustainable Tourism, 20*(3), 319–338.

Kowal, E. (2008). The politics of the gap: Indigenous Australians, liberal multiculturalism, and the end of the self-determination era. *American Anthropologist, 110*(3), 338–348.

Laverack, G., & Thangphet, S. (2007). Building community capacity for locally managed ecotourism in Northern Thailand. *Community Development Journal, 44*(2), 172–185.

Leach, M., Mearns, R., & Scoones, I. (1999). Environmental entitlements: Dynamics and institutions in community-based natural resource management. *World Development, 27*(2), 225–247.

Novelli, M., & Tisch-Rottsteiner, A. (2012). Authenticity versus development: Tourism to the hill tribes of Thailand. In O. Moufakkir, & P. M. Burns (Eds.), *Controversies in tourism* (pp. 54–72). Oxon, UK: CABI.

Palmer, N., & Chuamuangphan, N. (2017). Societal values and local responses to appropriate livelihoods, fairness and decision-making involvement in ecotourism: Chiang Rai, Thailand. *Asian Journal of Tourism Research, 2*(1), 118–149.

Roe, D., Goodwin, H., & Ashley, C. (2004). 1 2 Pro-poor tourism: Benefiting the poor. *New Horizons in Tourism: Strange Experiences and Stranger Practices, 147*, 1–4.

Scheyvens, R. (1999). Ecotourism and the empowerment of local communities. *Tourism Management, 20*(2), 245–249.

Scheyvens, R. (2002). *Tourism for development: Empowering communities.* Essex: Pearson Education.

Scheyvens, R. (2012). *Tourism and poverty.* London: Routledge.

Shenton, A. K. (2004). Strategies for ensuring trustworthiness in qualitative research projects. *Education for Information, 22*(2), 63–75.

Stonich, S. C. (1998). Political ecology of tourism. *Annals of Tourism Research, 25*(1), 25–54.

Tosun, C. (2000). Limits to community participation in the tourism development process in developing countries. *Tourism Management, 21*(6), 613–633.

UNCHR. (2015). *The UN refugee agency 2015 country operations profile.* Retrieved from http://www.unhcr.org/pages/49e489646.html

UNDP. (2014). *Thailand human development report*, 2014. Retrieved from http://hdr.undp.org/en/content/advancing-human-development-through-asean-community

UN Statistics Division. (2013). Retrieved from http://unstats.un.org/unsd/demographic/meetings/egm/Thailand/2013/agenda.pdf

Walpole, M. J., & Goodwin, H. J. (2001). Local attitudes towards conservation and tourism around Komodo National Park, Indonesia. *Environmental Conservation, 28*(02), 160–166.

Welcome to Chiangmai and Chiangrai. (2014). *The Thai village.* Retrieved from http://www.chiangmai-chiangrai.com/the_thai_village.html

Yin, R. K. (2003). *Designing case studies.* London: Sage.

Sustainable ecotourism management and visitor experiences: managing conflicting perspectives in Rocky Mountain National Park, USA

Tek B. Dangi and William J. Gribb

ABSTRACT

Management of tourism attractions, including national parks, emphasizes partnership and collaboration among various stakeholders for resource sustainability and visitor satisfaction. Balancing visitor experiences, park resources, and concessionaires' support for park policy and programs, however, has always been challenging. Past studies have attempted to address issues related to park management/stakeholders, collaboration, partnership, and governance, yet few have examined how differing perspectives of management and service providers (e.g. concessionaires) impact sustainable ecotourism management and visitor experiences. The purpose of this research was to explore perceptions (both similar and dissimilar) of two key stakeholders – park managers and concessionaires inside the park – at Rocky Mountain National Park (RMNP), U.S.A., regarding sustainable ecotourism management and visitor experiences. Twenty-seven park concessionaires and nine management staff responded to a brief questionnaire. Results showed varying areas of agreement and disagreement. While both groups agreed on ecotourism activities and their overall positive impacts, opinions differed on carrying capacity, horse use impact, and visitor conflict issues. Analyzing and discussing these differences from the theoretical perspectives of stakeholder and collaboration theories, the research carries management implications as well as a need for additional research.

1. Introduction

Demand for visitation of national parks and publicly managed lands (protected areas) have been growing significantly. National parks and other managed public lands have attracted tourists and recreationists for their natural, esthetic, educational, historic, scientific, and recreational qualities. National parks provide significant social and economic contributions to the local, regional and national economy, in addition to their role in preserving natural resources (Ferreira & Harmse, 2014; NPS, 2015a; O'Brien, 1999; Runte, 1987).

Against the backdrop of the popularity of protected area (hereafter referred to as national park) experiences, increasing demand for park visitation and resultant need for services and facilities has posed new planning and management challenges. Development of park infrastructure, services, and facilities have often resulted into negative impacts on natural and cultural resources and on visitor enjoyment (Ferreira & Harmse, 2014; Newsome, Milewski, Phillips, & Annear, 2002; Wilson, Nielsen, & Buultjens, 2009). Two critical objectives of national parks are to attain resource sustainability while providing the public with enjoyment (NPS, 2015a). Preserving the attractive qualities of national parks has been impeded, however, by factors such as air and water pollution, congestion, overcrowding, development impacts on the periphery, unethical business practices, and stakeholder conflicts (Freemuth, 1991; Jamal & Stronza, 2009; McCool, 2009; Plummer & Fennell, 2009; Sax, 1980; Tourtellot, 2005). Barriers exist to integrated sustainable tourism development in the form of exogenous obstacles, such as political influence and interventions, conflict of interest among various stakeholders, and ineffective regulations (Soteriou & Coccossis, 2010). Amidst such challenges, sustainable tourism development has gained momentum in theory and practice, especially in the management of national parks.

Sustainable development has been defined as 'development that meets the needs of the present without compromising the ability of future generations to meet their own needs' (WCED, 1987, p. 8). This definition has served as a mantra for sustainable development in various sectors, including tourism. Researchers suggest that 'sustainable tourism developed alongside, but separate to, its parent paradigm of sustainable development' (Ruhanen, Weiler, Moyle, & McLennan, 2015, p. 518). Consequently, sustainable tourism development became manifested in various forms: Buckley (2012) concluded sustainable tourism forms include responsible tourism, ecotourism, conservation tourism and community tourism. Ecotourism originated back in the 1980s in the wake of the environmental movement and nature conservation for countering negative impacts of mass tourism. In the course of development, however, ecotourism refined itself as a practice of 'sustainable tourism in natural areas' (Pforr, 2001, p. 70). It has been argued that the overarching principles of social, economic and environmental sustainability closely integrate concepts such as sustainable development, sustainable tourism and ecotourism (Bramwell, 2007; Hall & Lew, 1998; Pforr, 2001; UNWTO, 2015). In a conceptual overview of ecotourism, Diamantis (2004) broadened the guiding principles of ecotourism to include that it… 'occurs in natural settings (protected and non-protected) with an attempt to increase benefits to the economy, society and environment through sustainable educational practices from locals to tourists and vice-versa' (p. 5). This is how historically ecotourism is placed as a synonym or an integral form of sustainable tourism. More specifically, within the broader framework of sustainable tourism, ecotourism has been defined as responsible travel to natural areas, which takes into consideration environmental conservation and wellbeing of the local people and provides learning opportunities (Donohoe & Needham, 2006; Fennell & Weaver, 2005; Hall & Lew, 1998; TIES, 2015).

1.1. Some areas of conflicts in the management of national parks and the context of the study

Undoubtedly, national parks are major playgrounds of ecotourism activities. However, increased park visitation and external impacts, coupled with internal issues of resource constraints, have posed sustainable ecotourism management challenges for national parks in the U.S. and elsewhere (Jamal & Stronza, 2009; Runte, 1987; Tourtellot, 2005). Moreover, issues of multi-stakeholder conflicts in relation to balancing park resources with visitor satisfaction have drawn attention for over three decades. Among such conflicts include those seen between 'environmentalists' and 'developers' in Banff National Park, Canada (Eyre & Jamal,1998), jurisdictional conflicts of interest among U.S. Federal agencies (such as the U.S. Forest Service and National Park Service) and value-based environmental agencies and local residents in the greater Yellowstone ecosystem (Goldstein, 1992), and between winter snowmobile-users and environmentalists in Yellowstone National Park (NPS, 2015b) have been well-documented in tourism research. Tourism businesses located within national parks may have limited opportunities for success due to limited participation in final management decisions and due to conflicts between their business interests and traditional goals and cultural values of management (Haukeland, 2011). Among the various activities undertaken in national parks, management of conflict over nature conservation and recreational demand for horseback riding has also been identified as a common problem in many national parks, especially in North America and Australia (Newsome et al., 2002). In the 1990s, political pressure to make Kruger National Park (South Africa) more accessible to the local population to ensure economic benefits posed some challenges to carrying capacity and sustainability (Ferreira & Harmse, 2014). Conflict and agreement in residents' and hotel managers' views of tourism impacts and forest-related tourism development have been reported in Turkey (Kuvan & Akan, 2012). Stakeholder conflicts pose serious management challenges which may also endanger the sustainable character of national parks. After presenting this short review of global and national trends and perspectives of conflicting issues among various stakeholders in national parks, this study proposes to focus around managing conflicting relations between the park concessionaires and management in Rocky Mountain National Park (RMNP) for sustainable ecotourism development. Several internal reports including *Final Master Plan* (1976), *RMNP Transportation Study* (2000) and external reports such as *National Geographic Traveler Special Report* (NGTSR; Tourtellot, 2005) have identified challenges relating to resource conservation and maintaining visitors' satisfaction in RMNP described in further detail in the relevant section.

Various theoretical and management approaches have been applied to address conflicting issues surrounding stakeholders, resource conservation, and visitor satisfaction in the context of national parks and for turning those challenges into opportunities. Some of these approaches include the rational-comprehensive model, complex system theory model, and partnership/collaboration approaches (Plummer & Fennell, 2009). Other scholars have applied post-disciplinary and multi-theoretical approaches such as social representation, network, environmental dispute resolution, organizational theory, and public policy research (Laing, Lee, Moore, Wegner, & Weiler, 2009; Wilson et al., 2009) to study public-private partnerships in protected areas. This research, however, focused on a partnership/collaborative approach of RMNP management drawing from stakeholder theory and collaboration theory. Tourism scholars believe that 'two distinct areas of thinking have emerged' (Byrd, 2007, p. 9) in tourism regarding stakeholder theory. The first is related to classical theory of stakeholder management in organizational contexts as per the definition, power, and influence of stakeholders (Bramwell & Sharman,

1999; Donaldson & Preston, 1995; Freeman, 1984). The second area that has been applied in sustainable community-tourism planning and protected area management over last three decades has been practiced as collaborative thinking and action (Byrd, 2007; Gray, 1985; Jamal & Getz, 1995; Gray & Wood, 1991; Jamal & Stronza, 2009; Waligo, Clarke, & Hawkins, 2013).

1.2. Context of stakeholder theory and collaboration theory for sustainable tourism

1.2.1. Stakeholder theory

Many management scholars including Kaler (2004) and Laplume, Sonpar, and Litz (2008) credit Freeman (1984) as the father of stakeholder theory who defined stakeholder as 'any group or individual who can affect or is affected by the achievement of the organization's objectives' (Freeman, 1984, p. 46). Stakeholders have also been defined as 'those groups without whose support the organization would cease to exist' (SRI, 1663 as cited in Freeman, 1984, p. 31): this highlights the significant role stakeholders play in any organization. Management scholars mainly classified stakeholders into two: internal (primary) and external (secondary) stakeholders which further incorporates a wide range of stakeholders such as owners, customers, employees, suppliers, governments, competitors, consumer advocates, communities, stockholders, environmentalists, special interest groups, the media and so on (Buchholz & Rosenthal, 2005; Kaler, 2009; Laplume et al., 2008). These definitions suggest stakeholders are individuals or groups who have some stake in an organization and in return, can be affected by organizational actions or inactions. Contrary to the traditional management focus on internal stakeholders, stakeholder theory takes into account all stakeholders. Mitchell, Agle, and Wood (1997) defined the role of stakeholders in terms of three fundamental attributes: power, legitimacy, and urgency and a player without possessing one was not considered as a stakeholder.

For its strengths such as bridging stakeholder relationships, building trust and relationship through collective decisions, using employee stock option programs, taking multi-stakeholder perspectives, employing direct stakeholder engagement, and making balanced, ethical decisions (Donaldson & Preston, 1995; Laplume et al., 2008), the stakeholder theory has been found highly relevant in sustainable tourism partnerships (Byrd, 2007; Waligo et al., 2013).

1.2.2. Collaboration theory

Collaboration and partnership in sustainable tourism/ecotourism is comprised of various stakeholders, such as the tourism industry, visitors/tourists (present and future), host communities (present and future), public sector/government, voluntary sector organizations, media, educational institutions, and the environment (Berno & Bricker, 2001; Bramwell & Lane, 1993; Byrd, 2007; Waligo et al., 2013). It has been adopted as a tool in many countries for the sustainable management of national parks (Moore & Weiler, 2009; Phillips, 2003). Wood and Gray (1991) define collaboration as 'Collaboration occurs when a group of autonomous stakeholders of a problem domain engage in an interactive process, using shared rules, norms, and structures, to act or decide on issues related to that domain' (p. 146). From environmental communications perspectives, Cox (2010) defined collaboration as constructive, open, civil communication, generally a dialogue, with a focus on the future, with an emphasis on learning, and some degree of power sharing. Collaboration has been defined as a process of managing differences constructively, working together for addressing the inter-organizational problems through joint decision, a sharing of power and resources, and respecting each participant's interdependence.

Similar to stakeholder theory, collaboration theory holds multiple stakeholder benefits: potential to increase resources, enhance management, and introduce innovation (Phillips, 2003); promote technology transfer, boost international competitiveness, collaborate through self-regulations, and forge public-private partnership on issues difficult to be resolved alone (Gray & Wood, 1991; McCool, 2009). Additional benefits include building consensus, resource sharing, collective decision, and accountability in the problem domain (Gray, 1985; Jamal & Getz, 1995; Jamal & Stronza, 2009; Pforr, 2006); an opportunity to improve policy coordination (Bramwell & Sharman, 1999); and to enhance quality and legitimacy of the decisions, and to reduce lawsuits (Byrd, 2007). Scholars have argued for collaboration and adaptive co-management as a tool for forging better stakeholder understanding and partnership as it applies cross-disciplinary knowledge and information to tackle the issues of complexity, change and uncertainty relating to the management of national parks in the twenty-first century (Jamal & Stronza, 2009; McCool, 2009; Plummer & Fennell, 2009).

There are many common elements in stakeholder theory and collaboration theory, such as achieving organizational objectives through individuals and/or groups of stakeholders, pooling of tangible/intangible resources, and common problem-solving approach. Wood and Gray (1991) argued that the purpose of a single theoretical perspective or a general theory of collaboration was yet to be achieved. Hence, stakeholder theory and collaboration theory were deemed appropriate to analyze and discuss the issues in this study as they proffer an elaborate roadmap on collaboration and partnership success among stakeholders applicable to the national park management as well. Moreover, conflicting perceptions of the management and concessionaire specifically in RMNP were rarely addressed in literature as impeding stakeholder relationship or management issues. Insights and guidance taken from the theoretical perspectives of stakeholder and collaboration theory suggested that some of the issues and conflicts in RMNP management relating to sustainable ecotourism were not beyond solutions. Therefore, this research investigated why such differences occurred and what would be the way forward.

1.3. Study site: Rocky Mountain National Park and sustainable ecotourism

RMNP lies in the great Rocky Mountain chain that extends 4345.23 km from Mexico to Alaska. Located in Colorado, the park boasts over 2133.6 m in elevation change, and visitors can experience a breathtaking journey. The park has 72 mountain peaks higher than 3657.6 m (RMNP, 2008). The RMNP also bears the legacy of 10,000 years of human history, as Paleo-Indians and Native Americans are said to have used and crossed today's Trail Ridge for game hunting (OhRanger.com, 2015). The park's Trail Ridge Road (TRR) – the highest paved continuous highway in the U.S. – offers an exciting diversity of experience between alpine, subalpine and montane climate and flora and fauna, all available at one location.

Celebrating its centennial year in 2015, RMNP is one of the most visited national parks in the United States. RMNP was established by an act of Congress on January 26, 1915:

> … and said tract is dedicated and set apart as a public park for the benefit and enjoyment of the people of the United States … with regulations being primarily aimed at the freest use of the said park for recreation purposes by the public and for the preservation of the natural conditions and scenic beauties thereof. (RMNP, 1976, p. 1)

The park's *Final Master Plan* (1976) stated that the park experience of visitors should be a dynamic and vibrant one, but not at the cost of perpetuation of natural processes. A large challenge for RMNP, however, is to perpetuate dynamic ecosystems and sustain its natural forces. The *Final Master Plan* includes specific directions regarding sustainability: 'restore missing plant, animal, and fish types, and ensure that natural environmental rhythms continue' (p. 14).

Despite such directions, almost three decades after the *Final Master Plan* was in place, the *RMNP Transportation Study* (2000) also suggested that the RMNP faced problems of traffic congestion, overflow parking, pedestrian and motorist safety. In the NGTSR, *Destination Scorecard* study of 55 national parks in Canada and U.S.A. on a 1–100 scale (from near-pristine to near death), RMNP fell in the 'facing trouble' category.[1] The RMNP score of 56 was given stating some reasons: its 'deep blue glacial lakes and majestic mountain views.' in a 'totally overstressed' park. 'Traffic, condo developments, and noise around the sappy, touristy town of Estes Park lessen appeal' (Tourtellot, 2005, p. 90). The RMNP's 'facing trouble' rating confirmed the *Transportation Study*'s findings of traffic congestion and identified additional issues of condo development and noise pollution, which compromise the appeal of the park. These reports suggest that the natural environment rhythm of RMNP, which the *Final Master Plan* envisioned, has been questioned.

The *Final Master Plan* also considered ever-increasing visitor numbers and their modern comfort standards as a threat to the existing park resources, a finding seconded by Tourtellot's (2005) study. In summer, roads and trails become overcrowded. Conflicts between horseback riders and hikers are common during busy hours. The alpine tundra has been identified as highly fragile and sensitive to human activities. Congestion is taken as such a challenge that management proposed to reduce congestion in the Hidden Valley by establishing carrying capacity scheme or allowing weeklong uses (RMNP, 1976). The park management introduced a backcountry permit system in 1972 to address some of the congestion issues. Continuing growth of residential and commercial areas encircling the park has been cited as another challenge to sustain the environmental integrity and authenticity of the area (RMNP, 1976).

The Trail Ridge Store and adjacent visitor center have been identified as the reason for vehicle congestion in the Park's fragile and sensitive alpine environment. Besides prolonged traffic congestion owing to the presence of the store and visitor center, additional issues of sanitation, parking, and water supply conditions are exacerbated in such an eco-sensitive environment. Park management proposed to reverse the area's overdevelopment to decongest it in the long run. The *Final Master Plan* also underlined the need to reduce horse use in the areas of heavy use, near stables and take-off points, on steep grades and on dead-end trails. Though horse use in the park has been historic and desirable, it must be balanced with other activities to minimize its impact. The plan also proposed to remove two interior stables by 1979 through a study of merits and disadvantages. The possible solutions proposed in the *Final Master Plan* to address hiker-horse conflict included dual trail systems, preferential zoning for both uses, and improved trail circulation systems.

Besides the independent study by NGTSR (Tourtellot, 2005), several internal studies, including the *Transportation Study* highlighted the problem of congestion and overcrowding. The *Annual Narrative Report of Superintendents* (RMNP, 2006) indicated issues of high atmospheric deposition of sulfur and nitrogen in the park, including additional problems of bark beetle and visual impairment (air pollution), and a publication at a recent RMNP Research Conference (Davis, 2012) drew attention to visitors' concerns on anthropogenic climate change impacts for RMNP. A growing challenge for the park is to perpetuate dynamic ecosystems and its natural forces, which are further at risk due to the growing number of visitors every year – 3.4 million in 2014 (RMNP, 2015). The *Final Master Plan* asserted that 'there is an increasing impact on the environment as measured by the effects on vegetation and wildlife … [and] there is a growing impact on the experience, as demonstrated by crowding and conflicts in lifestyle' (p. 10).

Conserving park natural resources while maintaining and enhancing visitor satisfaction seemed a daunting task for RMNP management alone without forging effective collaboration with other park stakeholders. This research aimed to investigate the perspectives of both management and park concessionaires pertaining to sustainable ecotourism management of RMNP. Extensive work has been done on sustainable tourism, ecotourism, and developing collaboration for sustainable tourism development, but less focused research has been undertaken to find differences and similarities of perceptions between the management and concessionaires for sustainable ecotourism management in a national park in the United States. Considering a dearth of literature in delineating the differences of perspectives between management and concessionaires, this study explores such similarities and differences and suggests alternatives to improve collaboration and to maintain the sustainable ecotourism character of the RMNP applying stakeholder and collaboration theory. To help investigate these issues and identify similarities and differences between management and concessionaire perspectives, the main research questions included:

(1) How do concessionaires and park managers view/define ecotourism and relate it in the context of RMNP?
(2) What are the major challenges of RMNP relating to ecotourism and sustainable development and enhancing visitors' satisfaction?
(3) What are the suggestions and recommendations of concessionaires and management to address ecotourism and sustainable development issues they identify?
(4) Do the management and concessionaires have similar or different opinions to address ecotourism and sustainable development issues they identify?

2. Methods

The study area was specifically focused on the management strategy of RMNP in relation to sustainable ecotourism development and management vs. concessionaires' perceptions on the same. The study method specifically consisted of two separate questionnaires, one for park management and another for concessionaires. Key park officials and park concessionaires were targeted as part of a non-probability (purposive) sampling technique (Bernard, 2013).

After park management approval was obtained, the park concessionaires' questionnaire was circulated to all 114 park concessionaires and completed by October 30, 2009. Both electronic and printed questionnaires were sent, with return envelopes welcoming response either by mail or e-mail. Twenty-seven responses were received (23.7%). The park management questionnaire was circulated electronically to all 74 park management staff (supervisors and work leaders) by the RMNP supervisory management specialist of business programs. Responses from park management staff were completed by May 19, 2010, with a response rate of 12.2% (nine responses).

The questionnaires for RMNP management and concessionaires each contained 16 questions, starting with a brief definition of ecotourism and sustainable tourism in order to relate the respondent to sustainable ecotourism development in RMNP. Questions were of mixed type, open-ended for suggestions and closed-ended following five-point Likert Scale format ranging from 'strongly disagree' to 'strongly agree' with a column for 'not applicable' (NA) when necessary. Open-ended questions were designed to generate explanatory and qualitative responses to enhance the understanding of the study focus. The Likert scale questions were assigned to elicit participants' perceptions on certain issues rather than producing a comprehensive statistical analysis.

2.1. Profile of respondents

From the 27 concessionaires, the majority of the responses came from those involved in guided hiking and backpacking ($n = 10$), followed by fly fishing instruction, photography, studio, art/photo workshop, camps, and other service providers (such as shuttle services, park lodges, and gateway stables). Of the 27 concessionaires, 15 (56%) have had more than 10 years of operational experience, while the remainder had less than 10 years. Twenty operators (74%) handled less than 500 clients annually, three operators handled between 501 and 1000 clients, one operator between 1501 and 2000 clients, and two operators above 2001 clients annually. RMNP respondents included nine management-level officials, including managers, research administrators, interpreters, biologists, supervisors, and management specialists who had an average of 8.6 years and 12. 8 years of experience in RMNP and NPS, respectively.

3. Analysis of results

3.1. Views on ecotourism and related activities

3.1.1. Meaning of ecotourism

Both the concessionaires and management were asked to define ecotourism in the context of RMNP (Table 1). Based on the responses received, however, management tends to place more focus on educating tourists about the place, activities that are environment-friendly, natural settings, nature conservation, and sustainable development, whereas relaxation and culture conservation received low priority from managers. The concessionaires' focus, however, was more inclined to esthetic features such as natural settings, beauty of the mountains, activities that are environment-friendly, nature conservation, and educating tourists about the place among top five, whereas culture conservation and benefitting the local community received low rankings. Surprisingly, relaxation was not

Table 1. Management and concessionaires' definitions of ecotourism in the context of RMNP.

	% of concessionaires ($n = 27$)	% of management ($n = 9$)
Natural settings	92.6	77.8
Beauty of the mountain	92.6	44.4
Nature conservation	88.9	77.8
Culture conservation	29.6	33.3
Activities that are environment-friendly	92.6	88.9
Benefiting local community	51.9	66.6
Educating tourists about the place	81.5	100.0
Relaxation	63.0	33.3
Sustainable development	51.9	77.8
Other*	N/A	N/A

*There were a limited responses in other where concessionaires' responses included 'enjoying God's creation,' and 'teaching about it,' while managers' responses included 'protecting natural quiet,' 'clean air and water,' and 'night sky w/no or little artificial light.'

important for management (in contrast with NPS objectives), which is higher for concessionaires.

3.1.2. Major ecotourism activities

Among the list of the top twelve activities undertaken by their clients, concessionaires and management responded that hiking/backpacking was the number one activity, followed by horseback riding, fishing, watching scenery by car, camping, climbing and mountaineering, rock climbing, bicycling, snowshoeing/snowmobiling, skiing and swimming. Six respondents added learning photography, wildlife photography, family photography, and landscape photography in the preferred list of activities by their clients. The common understanding of major tourism activities seemed to form a common ground for future planning.

3.2. Major sustainable ecotourism challenges to RMNP

3.2.1. Options for reducing congestion and enhancing visitor experiences

The questionnaire also sought to receive feedback on two proposed options to reduce congestion and enhance the visitor experience: (1) change location of souvenir shop; and (2) develop shuttle system for TRR and Old Fall River Road (OFRR). The question sought feedback, given the two options, regarding which step would better contribute to reducing congestion, enhancing business, and visitor experience in the Alpine Visitor Center. The majority of park operators believed a change in location for the souvenir shop would not solve congestion problems, which the management closely agreed with (Table 2). Almost half of concessionaires agreed that developing a shuttle system would help solve congestion problems, while the majority of management also agreed this option would enhance the visitor experience (Table 2).

Regarding reduction of congestion along TRR or at the Alpine Visitor Center, five management staff made additional suggestions which included implementing an intelligent transportation system and to encourage alternative destinations/patterns of use as well as encourage alternative scheduling to diffuse temporal concentrations of visitor use patterns. Further suggestions included to change the location of the souvenir shop, to develop a shuttle system for TRR and OFRR, to arrange more formal paved pullouts, to get rid of souvenir shop, and to provide similar experience at another location.

Table 2. Issues/topics where management and concessionaires agreed or strongly agreed.

Management ($n = 9$)							Concessionaires (n varies)*					
Strongly Disagree %	Disagree %	Neutral %	Agree %	Strongly Agree %	Not Applicable %	Issues/topics of concern	Strongly Disagree %	Disagree %	Neutral %	Agree %	Strongly Agree %	Not Applicable %
22.2	44.4	0.0	33.3	0.0	0.0	*Change location of souvenir shop*[1]	26.9	19.2	15.4	3.8	23.0	11.5
0.0	22.2	0.0	44.4	33.3	0.0	*Develop shuttle system for TRR and OFRR*[2]	22.2	11.1	11.1	7.4	37.0	11.1
						Economic impacts of tourism[1]						
0.0	0.0	0.0	33.3	66.6	0.0	Creates jobs for local residents	7.7	3.8	0.0	38.5	50.0	0.0
0.0	0.0	0.0	44.4	55.6	0.0	Increases community income	3.8	3.8	7.7	30.8	53.8	0.0
0.0	0.0	11.1	44.4	44.4	0.0	Stimulates local commerce	3.8	3.8	7.7	30.8	53.8	0.0
0.0	0.0	0.0	55.6	44.4	0.0	Increases local govt. revenues	3.8	7.7	3.8	30.8	53.8	0.0
0.0	22.2	0.0	44.4	33.3	0.0	Spreads overall development	0.0	15.4	30.8	23.0	30.8	0.0

*Varying number of concessionaire responses depending on question/s.
[1]$n = 26$ for concessionaires.
[2]$n = 27$ for concessionaires.

3.2.2. Social, economic, and environmental impacts of tourism in the park

Regarding economic, environmental, and social impacts of tourism in the local community, both management and concessionaires (Table 2) were optimistic about the positive impacts. Concessionaires and management mainly 'agreed' or 'strongly agreed' with the positive economic contribution of tourism to the community. Regarding environmental impacts of tourism, concessionaires expressed their confidence on the positive environmental contribution of tourism in terms of conservation awareness and other impacts as majority of them either 'agreed' or 'strongly agreed.' In the case of park management, however, the result was rather mixed, as 50% of respondents remained neutral and 50% 'disagreed' that products remained compatible to the local environment. Moreover, management 'disagreed' that more tourism enhanced product awareness. There was encouraging response both from the management and operators on the positive social impacts of tourism, as the majority of respondents either 'disagreed' or 'strongly disagreed' that tourism had negative social impacts such as increased crime rates or degraded cultural environment.

3.2.3. Visitor compliance and cooperation

Visitors' compliance and cooperation is considered a prerequisite for the success of sustainable tourism businesses including national park settings. Damage and destruction of resources worth millions of annual reinvestments have been reported due to non-compliance (Johnson & Kamp, 1996). Regarding the rate of compliance by visitors for park resources use, some respondents (Table 3) expressed views that RMNP was much better than other parks due to visitor compliance. Moreover, 70.4% of concessionaires found visitors' compliance good or excellent, and 88.9% of management staff found visitors' compliance good, which is a strong indicator of park sustainability from visitors' compliance perspective.

3.3. Differences in perspectives

3.3.1. Impacts of horseback riding

The impact of horseback riding was one of the major concerns to sustainability. A majority of concessionaires felt that horseback riding was within limits as stated in the *Master Plan* (Table 3). One respondent felt that RMNP was not as significantly affected as other parks by horseback riding. A slightly modified question of horseback riding impacts was asked of managers: they were asked to rank the impacts of horseback riding on the park (Table 4). The majority (nearly 60%) of management either 'strongly disagreed' or 'disagreed' that horseback riding was within the limits in the park, especially on take-off points, on steep grades or dead-end trails.

Table 3. Management and concessionaires' perceptions of visitors' compliance and cooperation in RMNP.

	Excellent %	Good %	Average %	Poor %	Horrible %
Concessionaires ($n = 27$)	14.8	55.6	26.0	3.7	0.0
Management ($n = 9$)	0.0	88.9	0.0	11.1	0.0

Table 4. Issues/topics where management and concessionaires disagreed or strongly disagreed.

Management (*n* varies)*					Issues/topics of concern	Concessionaires (*n* varies)*				
Strongly Disagree	Disagree	Neutral	Agree	Strongly Agree		Strongly Disagree	Disagree	Neutral	Agree	Strongly Agree
%	%	%	%	%		%	%	%	%	%
–	–	–	–	–	*Horse-back riding impacts within limits*[1]	3.8	19.2	23.0	38.5	15.4
11.1	33.3	33.3	22.2	0.0	*Horse-back riding impacts*[2]	–	–	–	–	–
					Is within limits at take-off points					
11.1	55.6	22.2	11.1	0.0	Is within limits on steep grades	–	–	–	–	–
11.1	55.6	33.3	0	0.0	Is within limits at dead-end trails	–	–	–	–	–
–	–	–	–	–	*Park enjoyment without conflicts*[3]	7.7	11.5	11.5	46.2	23.0
11.1	55.6	22.2	11.1	0.0	*Park enjoyment without conflicts*[2]	–	–	–	–	–
					Horseback riders are affected					
0.0	11.1	0	33.3	55.6	*Hikers are affected*	–	–	–	–	–
0.0	44.4	11.1	44.4	0.0	*Carrying capacity, EIAs, and Plans*	–	–	–	–	–
					RMNP has adequate law enforcement[2]					
11.1	22.2	22.2	44.4	0.0	RMNP does not have adequate law enforcement[4]	–	–	–	–	–
12.5	12.5	25.0	50.0	0.0	RMNP's current EIAs and plans are adequate[4]	–	–	–	–	–
0.0	22.2	22.2	44.4	11.1	RMNP needs to develop new EIAs and plans[2]	–	–	–	–	–
0.0	0.0	20.0	40.0	40.0	Parking limits and quotas should be introduced in critically congested places[5]	–	–	–	–	–
–	–	–	–	–	*Carrying capacity (if limits & quotas imposed)*	14.3	14.3	21.4	21.4	28.6
					Horseback riding will be affected[6]					
–	–	–	–	–	Selling wood in campgrounds will be affected[7]	25.0	37.5	12.5	25.0	0.0
–	–	–	–	–	Selling wood/souvenirs will be affected[8]	20.0	30.0	10.0	40.0	0.0
–	–	–	–	–	Guiding, climbing will be affected[6]	21.4	14.3	14.3	14.3	35.7

*Varying number of management and concessionaire responses depending on question/s.
[1]$n = 26$ for concessionaires.
[2]$n = 9$ for managers.
[3]$n = 26$ for concessionaires.
[4]$n = 8$ for managers.
[5]$n = 5$ for managers.
[6]$n = 14$ for concessionaires.
[7]$n = 8$ for concessionaires.
[8]$n = 10$ for concessionaires.

3.3.2. Park enjoyment without conflicts (Among various park users)

Results indicated concessionaires felt confident that horseback riders, hikers, and bicyclists could enjoy a full ecotourism experience in the park without conflicts with other park visitors (Table 4). Most managers (89%) disagreed that horseback riders are affected, but felt that hikers were affected by other activities (Table 4). Moreover, management clarified that there were no trails open to bicycle use within the park, and visitors could use bicycles on park roads only, but horseback riding trails and hiking trails were not separated.

3.3.3. Carrying capacity concerns

When asked whether current law enforcement and plans of RMNP were adequate or whether new environmental impact assessments (EIAs), plans, or parking limits or quotas should be introduced in critically congested places, management's opinion was divided (Table 4). On the issue that RMNP did not have adequate law enforcement, almost half of management staff (44.4%) agreed with the notion, with an equal number of disagreement (44.4. %), and a small number (11.1%) who were neutral. A majority (50%) of management staff 'agreed' that RMNP's current EIAs and plans were adequate, as opposed to 25% who 'disagreed' and 25% were 'neutral.' A majority of managers (55.5%) recommended that RMNP develop new EIAs and plans, while almost one-fifth (22.2%) 'disagreed' and one-fifth (22.2%) were 'neutral' on this issue. Out of five responses received in the issue of whether parking limits and quotas should be introduced in critically congested places, four (80%) of the management either 'agreed' or 'strongly agreed' that quotas should be in place, which was the opposite opinion received from concessionaires.

Park concessionaires expressed concern with the carrying capacity limits and quotas which park management had imposed or would plan to impose in future in crowded/congested areas (Table 4), which included total annual visitor quota to the park, total number of horseback rides allowed, total number of camping parties allowed in a year/season, and total number of climbing/mountaineering members/teams allowed. Their concern regarded that any limits to be imposed in future would affect business. The majority of concessionaires involved in horseback riding, guiding climbing/mountaineering, fishing/fly fishing, general hiking, backpacking and photo workshops either 'agreed' or 'strongly agreed' that their businesses will be affected if quotas were enforced. Some concessionaires disagreed that selling wood in campgrounds, selling food/souvenirs, or backcountry use would be affected. Interestingly, close to 48.2% respondents (10–14 concessionaires in different categories) made no responses on these issue items.

3.4. Concessionaires and management suggestions to sustainable ecotourism

3.4.1. Concessionaires suggestions

In an open-ended question on the matter of strengthening partnership with the park management to contribute to ecotourism development, some concessionaires made a host of suggestions. The major suggestions included visitors should feel welcome, informing concessionaires about resource-conservation efforts, restoring the park ecosystem for wildlife, provision for additional park education, attendance of park concessionaires at regional recreation and community meetings, linking and listing of concessionaire websites as approved website, organizing more informative talks, improved wrangler/employee training from park personnel, and conducting more trail maintenance work.

3.4.2. Park management suggestions

When management was asked what they believed had been the most successful ecotourism and sustainable development practices adopted by the park, the responses could be grouped into four categories. The environmental practices included the use of bear canisters and waste disposal bags in the backcountry, limiting growth/business in the park, limiting dogs, bikes, and horses in the park, no hotels in the park, and energy friendly facilities. Educational/regulatory achievements included permits for guides/tourism groups, shuttles on TRR with ranger/interpretation; and ranger-led hiking/snowshoeing. Infrastructure/facility improvement activities included shuttle system and tie to town, providing multiple camping options, providing front country camping within the park, established backcountry sites and privies, good network of trails (well maintained), and visitor centers. Social activities included using park volunteers for activities in the local community.

3.4.3. Management views on best practices, concessionaire services, barriers, and collaboration techniques

Best practices (BP) in managing the park experience suggest that park and recreation agencies update their system master plan every 10 years, individual park master plans every 10 years, and that they maintain a yearly business plan for all core services the agency manages (Younger, 2009). Management responded that RMNP failed to have a new master plan every 10 years as per BP standards, but it has several sectoral plans which underline and implement the guidelines outlined in its *1976 Final Master Plan*. Some managers, however, underlined the need for a new master plan.

When management was asked what they thought were the major challenges and obstacles to RMNP for ecotourism and sustainable development in general and in relation to park concessionaires in particular, seven out of nine staff suggested better communication between park and concessionaires, training for concession employees, services needed to accommodate visitors (lodging, food, gas, etc.), that concession activities should be moved outside park boundaries wherever possible, and the ability for small businesses to achieve competitive and sustainable goals. Suggestions relating to challenges to park management included sheer number of visitors (also contributed by shuttles), parking issues, need to address issue of horses and horse waste, more action on feeding wildlife, balancing the needs of the people and partners with needs of the environment, and need of designated expert/s.

Another open-ended question asked the management to list the top five collaboration techniques they recommend to work with stakeholders. Six respondents made suggestions, including transportation planning (shuttle) with/to town, development of guidelines and restrictions on liveries, generating public comments in scoping meetings before an environmental impact statement, regular meetings, recycling programs (with town, Larimer County/Estes Park), development of nitrogen deposition reduction plan with the State of Colorado, and future multi-use trail system within the town and park.

4. Discussion

The purpose of the study was to explore the similarity and differences between perspectives of RMNP management and concessionaires and forward suggestions that hold implications for both the management and concessionaires relating to stakeholder and collaboration theory. A recent *Natural Resource Vital Sign* report of RMNP (Franke, Johnson, Ashton, & Bobowski, 2015, p. 1) also supported the findings from previous RMNP studies (Davis, 2012; RMNP, 1976; Tourtellot, 2005) that factors such as visitor congestion, climate change, and air and water pollution stood as the most significant natural resource issues affecting RMNP's ecosystem and attractions. The *Natural Resource Vital Sign* report supports this study's findings: RMNP management faces a host of challenges to maintain the ecosystem of the park in partnership with concessionaires. The report acknowledges the Clean Air Act requires RMNP to meet high air quality standards, 'however, atmospheric deposition, regional haze, and ozone formation are affecting park resources' (Franke et al., 2015, p. 8). Most RMNP visitors (78%) experienced some level of crowding, and visitors will continue to have a significant impact according to the report: 'as the human footprint continues to grow, so will its impact' (Franke et al., 2015, p. 5). It seems that studies such as this which recommended reducing congestion in critical areas for enhancing visitor satisfaction and ecotourism character of the park have helped inform the management to initiate appropriate actions. Recent research suggests that RMNP is considering a host of strategies to reduce crowding and negative impacts on park resources (Krzeminski, 2015). The strategies include informing and encouraging visitors to see other places in the park during high times and modifying the existing shuttle system to incorporate additional sites for similar experiences. RMNP states it is exploring the possibility of multi-use trail development on the east side of the park almost parallel to, but to be separate from existing road networks. The multi-use trail would be designed for non-motorized modes of trail-use and Environmental Assessment (EA) is underway to this effect (NPS, 2015c).

In the course of RMNP management directives following the RMNP's (1994) recommendations published in the *Commercial Horse Use Management Plan and Environmental Assessment*, two horse stables inside the park were not removed, but were allowed to continue operations (to date) with conditions on their use of horses and other pack animals in the designated trails. The RMNP allows horse-trails only in limited trailheads and horses are not permitted on all park roads excepting a few. In order to avoid conflict of horse-riders with hikers, NPS provides information and signage such as 'Heavy Horse Traffic' trails. Accommodation of horse stables inside the park and regulations for all types of recreational horse-use seems to respect visitor demand for horseback riding, considers concessionaires' business interest, and prioritizes environmental conservation. It follows some of the tips proffered in stakeholder and collaboration theory of taking multi-stakeholder perspectives, building trust and goodwill, sharing of resources, collective decision and accountability in problem domain (Gray, 1985; Jamal & Getz, 1995; Jamal & Stronza, 2009; Laplume et al., 2008; Pforr, 2006; Waligo, Clarke, & Hawkins, 2013). The possible solutions proposed in the *Final Master Plan* to address hiker-horse conflict through dual trail systems, preferential zoning for both uses, and improved trail circulation systems, however, are still yet to be implemented. The *Final Master Plan* solutions are consistent to Manning's (2011) recommendations of managing conflicts through zoning, which possibly could be an alternative to consider.

One of the interesting segments of this study finding was that despite the use of horses in the park as per the *Commercial Horse Use Management Plan* (RMNP, 1994), management expressed the opinion horseback riding was not within the limits on RMNP, hikers were affected due to other activities such as horseback riding, and horse stables should be relocated outside the park, which seem consistent with previous research at other parks in Australia (Newsome et al., 2002). The study found that recreational horse riding can cause a range of environmental impacts, including 'user-conflicts in multiple-use areas, loss of vegetation height and cover, change in plant species composition, degradation of existing trail networks' (p. 56). The Australian case study suggested management of various options, which included prohibiting use, unrestricted open access in some areas, and managing horse riding through commercial tour operators and so on. These recommendations made in Australian context can have implications on RMNP as the differing opinions (not on the regulations though) of the management and concessionaires regarding recreational horse use impacts portend potential tensions for the future. It may be time for RMNP management to address the issue of horse use impacts through a separate study in collaboration with concessionaires.

This study provides further evidence that differences occur among stakeholders due to values differences (Buckley, 2002; Lovelock, 2002). Many studies have agreed on the role of park management as the steward of park resources and for holding the long-term strategic vision rather than private businesses (Buckley, 2002; Haukeland, 2011; Kerstetter et al., 2010). This study identified a host of options that reflect both management's long-term vision and concessionaires' economic (business) interests. Some management practices such as limiting growth/business in the park, limiting dogs, bikes, and horses in the park and restrictions on park hotels speak of management's vision and action for conserving park resources for long-term use. However, RMNP concessionaires' voluntary activities such as partnership with 'Leave No Trace' (LNT) programs, hosting 'Green' weddings/events, public education of natural resources and so on deserve greater appreciation as they contribute to ecotourism operations. Findings of this study, however, differ slightly from Kuvan and Akan's (2012) study of residents' and hotel managers' views of tourism impacts and forest-related tourism development. Despite having some attitudes in common, significant differences were observed that management believed tourism had few environmental and social impacts where residents believed negative impacts were significant. Value differences being managers supported their business and employers, while local residents supported protection of the area.

The RMNP concessionaires made a host of suggestions to management regarding ecotourism development, including informing concessionaires about resource-conservation efforts, provisioning attendance of park concessionaires at regional recreation meetings, and organizing better wrangler/employee training from park personnel. These suggestions indicated some kind of information/communication gap or misunderstanding as suggested in stakeholder and collaboration theory. Laplume et al.'s (2008) findings suggest that stakeholders support organizations when they feel that they were being duly rewarded, considered important, and treated respectfully calls for consideration in this context. Incorporation of stakeholders in final management decisions can increase their participation, trust, work relationship, and spirit of collaboration (Haukeland, 2011; McCool, 2009).

5. Conclusion and recommendations

In order to understand the perspectives of management and concessionaires for sustaining park resources and enhancing visitor experiences for sustainable ecotourism in RMNP, the research asked four primary questions to the park management staff and concessionaires. Concessionaires and park managers were asked to define ecotourism in the context of RMNP, to identify the major challenges of RMNP relating to ecotourism and sustainable development and for enhancing visitors' satisfaction; provide suggestions and recommendations to address ecotourism and sustainable development issues, in order to determine whether park management and concessionaires had similar or different opinions.

Regarding sustainable ecotourism management in the park, there were some areas such as what contributes to ecotourism and associated activities, problems with the location of the souvenir shop/visitor center, use of shuttle bus system, positive social, economic impacts of tourism and visitor compliance to parks rules and code of conduct where both park management and concessionaires (outfitters) generally agreed. There seems to be a common understanding between the management and the concessionaires on what they viewed as RMNP's ecotourism attractions and activities such as hiking/backpacking, horseback riding, watching scenery by car, camping, climbing and mountaineering, and skiing. This common vision of tourism attractions greatly facilitates the responsibility of joint-marketing of RMNP as a tourist destination. There were some areas, however, where management and concessionaires had different perceptions, such as carrying capacity, impacts of horseback riding, and conflicts among various park visitors' activities.

Scholars (Buckley, 2002; Lovelock, 2002) find stakeholder tensions (conflicts) and differences healthy and natural, as long as a mechanism exists to respond. In the theoretical discussion of collaboration and stakeholder theory, multiple strategies have been suggested to address stakeholder conflicts. In the light of issues discussed and findings of this study, some of the steps suggested to park management to improve collaboration included: to further enhance communications with concessionaires, listen to stakeholder viewpoints, involve stakeholders in the decision-making process, reward stakeholders fairly, build trust and maintain relationships, take multi-party perspectives, structure for continued dependence upon each other, and understand stakeholders' value differences seriously (Cox, 2010; Donaldson & Preston, 1995; Gray, 1985; Jamal & Stronza, 2009; Laplume et al., 2008; McCool, 2009; Reynolds, Shultz, & Hekman, 2006; Waligo et al., 2013). One of the steps of rewarding stakeholders could be as simple as linking to and listing concessionaires' websites to identify them as recognized park partners.

The study explored the need of formulating a new, holistic master plan for RMNP replacing the 1976 *Final Master Plan*, incorporating several issues such as crowding and carrying capacity, especially differences of perceptions between management and concessionaires, and to extend partnership among several stakeholders. Bringing multi-stakeholders in the planning process and incorporating their interests, values, expertise, and suggestions can augment master plan formulation and implementation success. Farrell and Twining-Ward (2005) suggest tourism managers and planners understand sustainable tourism as complex system dynamics rather than equilibrium to be regained through traditional management practices of stability and predictability. A new master plan that embraces adaptive co-management and complexity could possibly provide better solutions for many existing and upcoming issues. Based on questionnaire feedback, *Final Master Plan* strategies, and best practices, this study suggests that a monitoring and evaluation procedure (unit) would be beneficial to RMNP for developing future sustainable ecotourism management strategies. Previous research supports the need for a monitoring and enforcement mechanism to reduce information gaps for continuing stakeholder relationships (Buckley, 2012; Hill & Jones, 1992).

The issue of transportation is important and can have long-term implications on park sustainability. Transportation planning with the town, the City of Estes Park, and introduction of shuttle system was believed to be a huge achievement by the management for bringing convenience to visitors and lessening congestion and environmental impacts. Some management staff argued, however, that the shuttle system had carried more passengers and contributed to additional congestion. Based on this finding, a critical departure time for RMNP has come as defined in the Director's Order #17 that 'high visitation numbers are not necessarily a desired goal of park management'(NPS, 2015d, p. 1). As outlined in Tourtellot's (2005) study, a critical time has approached for NPS to review its mission, which assigns to perform the dual but contrasting tasks of preserving unimpaired the natural and cultural resources for the enjoyment of its people, unlike the mandate given to Parks Canada to first protect the environment.

One limitation of this study was the low response rate, especially among park managers, and other stakeholders were intentionally excluded given the nature of study, and it would be dangerous to make statistical inferences from such low responses. Differences of perspectives between management and concessionaires on issues such as carrying capacity, horse use impacts, and visitor conflicts also warrant further cross-sectional and longitudinal study with inclusion of other major stakeholders. Collaboration is not a free gift: it sometimes increases organizational costs, reshapes structures and relationships, requires innovations, piles up additional responsibilities, and demands new skills development (Wood & Gray, 1991). Consideration of these holistic factors is suggested to further consolidate stakeholder collaboration and partnership in RMNP for sustainable ecotourism development.

Note

1. It is interesting to note that of the 55 parks in the National Geographic study, the majority of Canadian parks are in the top score category, the reason being 'By law, Parks Canada must first protect the environment, whereas Congress demands the U.S. National Park Service protect nature while also promoting outdoor recreation, dual mandates that can conflict when too many park-lovers show up' (Tourtellot, 2005, p.85).

Acknowledgements

The authors wish to thank the management of Rocky Mountain National Park (RMNP) for their authorization of the study, as well as their assistance with data collection efforts. The management opinions expressed herein are of the authors and do not represent the official viewpoint of RMNP management.

Disclosure statement

No potential conflict of interest was reported by the authors.

References

Bernard, H. R. (2013). *Social research methods: Qualitative and quantitative approaches*. California, CA: Sage Publications.

Berno, T., & Bricker, K. (2001). Sustainable tourism development: The long road from theory to practice. *International Journal of Economic Development*, *3*(3), 1–18.

Bramwell, B. (2007). Opening up new spaces in the sustainable tourism debate. *Tourism Recreation Research*, *32*(1), 19.

Bramwell, B., & Lane, B. (1993). Sustainable tourism: An evolving global approach. *Journal of Sustainable Tourism*, *1*(1), 1–5.

Bramwell, B., & Sharman, A. (1999). Collaboration in local tourism policymaking. *Annals of Tourism Research*, *26*(2), 392–415.

Buchholz, R. A., & Rosenthal, S. B. (2005). Toward a contemporary conceptual framework for stakeholder theory. *Journal of Business Ethics*, *58*(1–3), 137–148.

Buckley, R. (2002). Draft principles for tourism in protected areas. *Journal of Ecotourism*, *1*(1), 75–80.

Buckley, R. (2012). Sustainable tourism: Research and reality. *Annals of Tourism Research*, *30*(2), 528–546.

Byrd, E. T. (2007). Stakeholders in sustainable tourism development and their roles: Applying stakeholder theory to sustainable tourism development. *Tourism Review*, *62*(2), 6–13.

Cox, R. (2010). *Environmental communication and the public sphere*. California, CA: Sage publications.

Davis, S. (2012). *Visitor's concerns about climate change in ROCKY Mountain National Park*. Presented at 2012 Rocky Mountain National Park Conference, Estes Park, CO.

Diamantis, D. (2004). Ecotourism management: An overview. In Diamintis (Ed.), *Ecotourism: Management and assessment* (pp. 3–26). Hampshire: Cenage Learning EMEA.

Donaldson, T., & Preston, L. E. (1995). The stakeholder theory of the corporation: Concepts, evidence, and implications. *Academy of Management Review*, *20*(1), 65–91.

Donohoe, H. M., & Needham, R. D. (2006). Ecotourism: The evolving contemporary definition. *Journal of Ecotourism*, *5*(3), 192–210.

Eyre, M., & Jamal, T. B. (1998). Addressing stakeholder conflicts in a Canadian Mountain Park. *Parks and Recreation*, *33*(9), 87–94.

Farrell, B., & Twining-Ward, L. (2005). Seven steps towards sustainability: Tourism in the context of new knowledge. *Journal of Sustainable Tourism*, *13*(2), 109–122.

Fennell, D., & Weaver, D. (2005). The ecotourium concept and tourism-conservation symbiosis. *Journal of Sustainable Tourism*, *13*(4), 373–390.

Ferreira, S., & Harmse, A. (2014). Kruger National Park: Tourism development and issues around the management of large numbers of tourists. *Journal of Ecotourism*, *13*(1), 16–34.

Franke, M. A., Johnson, T. L., Ashton, I. W., & Bobowski, B. (2015). *Natural resource vital signs at Rocky Mountain National Park*. Natural Resource Report NPS/ROMO/NRR—2015/946. National Park Service, Fort Collins, Colorado.

Freeman, R. (1984). *Strategic management: A stakeholder approach*. Boston, MA: Pitman.

Freemuth, J. C. (1991). *Islands under siege: National parks and the politics of external threats*. Lawrence: University Press of Kansas.

Goldstein, B. (1992). The struggle over ecosystem management at Yellowstone. *BioScience*, *42*, 183–187.

Gray, B. (1985). Conditions facilitating interorganizational collaboration. *Human Relations*, *38*(10), 911–936.

Gray, B., & Wood, D. J. (1991). Collaborative alliances: Moving from practice to theory. *The Journal of Applied Behavioral Science*, *27*(1), 3–22.

Hall, C. M., & Lew, A. A. (1998). *Sustainable tourism: A geographical perspective*. Harlow: Longman.
Haukeland, J. V. (2011). Tourism stakeholders' perceptions of national park management in Norway. *Journal of Sustainable Tourism*, *19*(2), 133–153.
Hill, C. W., & Jones, T. M. (1992). Stakeholder-agency theory. *Journal of Management Studies*, *29* (2), 131–154.
The International Ecotourism Society (TIES). (2015, July). TIES Overview. Retrieved from http://www.ecotourism.org/ties-overview
Jamal, T. B., & Getz, D. (1995). Collaboration theory and community tourism planning. *Annals of Tourism Research*, *22*(1), 186–204.
Jamal, T., & Stronza, A. (2009). Collaboration theory and tourism practice in protected areas: Stakeholders, structuring and sustainability. *Journal of Sustainable Tourism*, *17*(2), 169–189.
Johnson, D. R., & Kamp, M. V. (1996). Extent and control of resource damage due to noncompliant visitor behavior: A case study from the US national parks. *Natural Areas Journal*, *16*(2), 134–141.
Kaler, J. (2004). Arriving at an acceptable formulation of stakeholder theory. *Business Ethics: A European Review*, *13*(1), 73–79.
Kaler, J. (2009). An optimally viable version of stakeholder theory. *Journal of Business Ethics*, *86*(3), 297–312.
Kerstetter, D. L., Mowen, A. J., Trauntvein, N. E., Graefe, A. R., Liechty, T., & Zielinski, K. D. (2010). Visitors' opinions of who should provide services and amenities in state parks. *Journal of Park and Recreation Administration*, *28*(4), 1–15.
Krzeminski, J. (2015). *Transportation and management strategies to reduce congestion in Bear Lake Road Corridor, Rocky Mountain National Park, USA*. Presented at 2015 Rocky Mountain National Park Conference, Estes Park, CO.
Kuvan, Y., & Akan, P. (2012). Conflict and agreement in stakeholder attitudes: Residents' and hotel managers' views of tourism impacts and forest-related tourism development. *Journal of Sustainable Tourism*, *20*(4), 571–584.
Laing, J. H., Lee, D., Moore, S. A., Wegner, A., & Weiler, B. (2009). Advancing conceptual understanding of partnerships between protected area agencies and the tourism industry: A postdisciplinary and multi-theoretical approach. *Journal of Sustainable Tourism*, *17*(2), 207–229.
Laplume, A. O., Sonpar, K., & Litz, R. A. (2008). Stakeholder theory: Reviewing a theory that moves us. *Journal of Management*, *34*(6), 1152–1189.
Lovelock, B. (2002). Why it's good to be bad: The role of conflict in contributing towards sustainable tourism in protected areas. *Journal of Sustainable Tourism*, *10*(1), 5–30.
Manning, R. E. (2011). *Studies in outdoor recreation: Search and research for satisfaction*. Corvallis, OR: Oregon State University Press.
McCool, S. F. (2009). Constructing partnerships for protected area tourism planning in an era of change and messiness. *Journal of Sustainable Tourism*, *17*(2), 133–148.
Mitchell, R. K., Agle, B. R., & Wood, D. J. (1997). Toward a theory of stakeholder identification and salience: Defining the principle of who and what really counts. *Academy of Management Review*, *22*(4), 853–886.
Moore, S. A., & Weiler, B. (2009). Tourism–protected area partnerships: Stoking the fires of innovation. *Journal of Sustainable Tourism*, *17*(2), 129–132.
National Park Service (NPS). (2015a, July). *About Us: The US Department of the Interior*. Retrieved from http://www.nps.gov/aboutus/index.htm
National Park Service (NPS). (2015b, June). *Yellowstone in Winter: Current management and planning*. Retrieved from http://www.nps.gov/yell/learn/management/currentmgmt.htm
National Park Service (NPS). (2015c, July). Multi-Use Trail Environmental Assessment. Retrieved from https://parkplanning.nps.gov/projectHome.cfm?projectID=45499
National Park Service (NPS). (2015d, July). Office of Policy. Retrieved from http://home.nps.gov/applications/npspolicy/DOrders.cfm
Newsome, D., Milewski, A., Phillips, N., & Annear, R. (2002). Effects of horse riding on national parks and other natural ecosystems in Australia: Implications for management. *Journal of Ecotourism*, *1*(1), 52–74.

O'Brien, B. R. (1999). *Our national parks and the search for sustainability*. Austin, TX: University of Texas Press.

OhRanger.com. (2015, July). *RMNP nature and wildlife in Rocky Mountain*. Retrieved from http://www.ohranger.com/rocky-mountain/toc/nature-wildlife

Pforr, C. (2001). Concepts of sustainable development, sustainable tourism, and ecotourism: Definitions, principles, and linkages. *Scandinavian Journal of Hospitality and Tourism, 1*(1), 68–71.

Pforr, C. (2006). Tourism policy in the making: An Australian network study. *Annals of Tourism Research, 33*(1), 87–108.

Phillips, A. (2003). Turning ideas on their head. In The George Wright Forum (Vol. 20, No. 2, p. 8).

Plummer, R., & Fennell, D. A. (2009). Managing protected areas for sustainable tourism: Prospects for adaptive co-management. *Journal of Sustainable Tourism, 17*(2), 149–168.

Reynolds, S. J., Shultz, F. C., & Hekman, G. R. (2006). Stakeholder theory and managerial decision-making: Constraints and implications of balancing stakeholder interests. *Journal of Business Ethics, 64*, 285–301.

Rocky Mountain National Park (RMNP). (1976). *Final master plan*. Estes Park, CO: RMNP.

Rocky Mountain National Park (RMNP). (1994). *Commercial horse use management plan and environmental assessment*. Estes Park, CO: RMNP.

Rocky Mountain National Park (RMNP). (2000). Transportation Study, Summary Report.

Rocky Mountain National Park (RMNP). (2006). *Annual narrative report of superintendents*. Estes Park, CO: RMNP.

Rocky Mountain National Park (RMNP). (2008). *Rocky mountain, backbone of the continent*. Estes Park, CO: RMNP.

Rocky Mountain National Park (RMNP). (2015, July). *Rocky Mountain, Facts and Figures*. Retrieved from http://www.nps.gov/romo/planyourvisit/upload/Facts-Figures.pdf

Ruhanen, L., Weiler, B., Moyle, B. D., & McLennan, C. L. J. (2015). Trends and patterns in sustainable tourism research: A 25-year bibliometric analysis. *Journal of Sustainable Tourism, 23*(4), 517–535.

Runte, A. (1987). *National parks: The American experience*. Lincoln, NE: University of Nebraska Press.

Sax, J. L. (1980). *Mountains without handrails: Reflections on the national parks*. Ann Arbor, MI: The University of Michigan Press.

Soteriou, E. C., & Coccossis, H. (2010). Integrating sustainability into the strategic planning of national tourism organizations. *Journal of Travel Research, 49*(2), 191–205.

Tourtellot, J. B. (2005, June). NGTSR Destination Scorecard. How do 55 national park regions rate? Retrieved from http://traveler.nationalgeographic.com/2005/07/NPDestinationScorecardJA05.pdf

United Nations World Tourism Organization (UNWTO). (2015, July). Sustainable Development of Tourism. Retrieved from http://sdt.unwto.org/content/about-us-5

Waligo, V. M., Clarke, J., & Hawkins, R. (2013). Implementing sustainable tourism: A multi-stakeholder involvement management framework. *Tourism Management, 36*, 342–353.

Wilson, E., Nielsen, N., & Buultjens, J. (2009). From lessees to partners: Exploring tourism public–private partnerships within the New South Wales national parks and wildlife service. *Journal of Sustainable Tourism, 17*(2), 269–285.

Wood, D. J., & Gray, B. (1991). Toward a comprehensive theory of collaboration. *The Journal of Applied Behavioral Science, 27*(2), 139–162.

World Commission on Environment and Development (WECD). (1987). *Our common future*. New York, NY: Oxford University Press.

Younger, L. (2009). *Best practices in managing high quality parks and park systems*. Presented at 2009 National Recreation & Park Association (NRPA) Annual Conference & Expo, Salt Lake City, UT.

Index